# Artificial Intelligence, Geographic Information Systems, and Multi-Criteria Decision-Making for Improving Sustainable Development

The Asia-Pacific region, home to some of the world's fastest-growing economies, faces a range of complex challenges, including environmental degradation, the increasing frequency of natural hazards, and rapid urbanization. Addressing these issues, which many countries across the globe are facing, requires innovative, interdisciplinary approaches to promote sustainable development and enhance resilience. Geographic information systems (GIS), when combined with multi-criteria decision-making (MCDM) techniques and advanced technologies such as artificial intelligence (AI), offer powerful tools to tackle these multifaceted problems. AIGIS integrates AI with GIS to derive insights from geospatial data. The fusion of AI techniques with GIS enhances data analysis, visualization, and decision-making. ***Artificial Intelligence, Geographic Information Systems, and Multi-Criteria Decision-Making for Improving Sustainable Development*** explore how these integrated tools can support decision-making processes aimed at advancing sustainable development.

Drawing on research and insights from diverse disciplines, the book looks at how GIS, MCDM, and AI can provide solutions for disaster risk reduction, environmental monitoring, urban planning, and natural resource management. Through diverse case studies and theoretical explorations, this book highlights the value of integrated geospatial tools in facilitating informed decision-making and fostering resilient societies in the face of evolving challenges. It covers a wide range of topics, including the following:

- Site-soil-geology assessments in Fiji
- Flood risk analysis in Hong Kong
- Air quality management in Delhi during the COVID-19 pandemic
- Vegetation health monitoring in Thailand

Bringing together the work of academicians, practitioners, and decision-makers, the book reflects the growing recognition towards effective and sustainable solutions to complex problems, which require a multidimensional approach, integrating scientific, economic, and social considerations. By providing the latest research and practical applications of MCDM, AI, and GIS, it contributes to ongoing efforts to build a more sustainable and resilient future for the Asia-Pacific region, as well as for the world.

**Advances in Computational Collective Intelligence**
*Series Editor Dr. Subhendu Kumar Pani*

*Deep Learning Concepts in Operations Research*
By Biswadip Basu Mallik, Gunjan Mukherjee, Rahul Kar, Aryan Chaudhary

*Applications of Machine Learning and Deep Learning on Biological Data*
By Faheem Masoodi, Mohammad Quasim, Syed Bukhari, Sarvottam Dixit, Shadab Alam

*Technologies for Sustainable Global Higher Education*
By Maria José Sousa, Andreia de Bem Machado, Gertrudes Aparecida Dandolini

*Artificial Intelligence Techniques in Power Systems Operations and Analysis*
By Nagendra Singh, Sitendra Tamrakar, Arvind Mewada, Sanjeev Kumar Gupta

*Deep Learning Applications in Operations Research*
By Aryan Chaudhary, Biswadip Basu Mallik, Gunjan Mukherjee, Rahul Kar

*Computational Intelligence in Industry 4.0 and 5.0 Applications: Trends, Challenges and Applications*
By Joseph Bamidele Awotunde, Kamalakanta Muduli, Biswajit Brahma

*Explainable Artificial Intelligence in Medical Imaging: Fundamentals and Applications*
By Amjad Rehman Khan, Tanzila Saba

*Parallel and High-Performance Computing in Artificial Intelligence*
By Mukesh Raghuwanshi, Pradnya Borkar, Rutvij H. Jhaveri, Roshani Raut

*Emotional Intelligence in the Digital Era: Concepts, Frameworks, and Applications*
By Pushan Kumar Dutta, Sachin Gupta, Shafali Kashyap, Anita Gehlot, Rita Karmakar, Pronaya Bhattacharya

*Deep Learning and Blockchain Technology for Smart and Sustainable Cities*
By Subramaniyaswamy V, G Revathy, Logesh Ravi, N. Thillaiarasu, Naresh Kshetri

*Computational Intelligence for Analysis of Trends in Industry 4.0 and 5.0*
By Joseph Bamidele Awotunde, Kamalakanta Muduli, and Biswajit Brahma

*Metaverse and Blockchain Use Cases and Applications*
By Dileep Kumar Murala, Sandeep Kumar Panda, and Sujata Priyambada Dash

*Leveraging Artificial Intelligence in Cloud, Edge, Fog and Mobile Computing*
By Shrikaant Kulkarni, P. William, Vijaya Prakash, and Jaiprakash Narain Dwivedi

*Using AI to Develop Sustainability Strategies for a Changing Global Economy*
By A.V. Senthil Kumar, Ankita Chaturvedi, Atul Bansal, Rohaya Latip

*Artificial Intelligence, Geographic Information Systems, and Multi-Criteria Decision-Making for Improving Sustainable Development*
By Sujoy Kumar Jana, Kamalakanta Muduli, Indrajit Pal, Purushottam Meena

https://www.routledge.com/Advances-in-Computational-Collective-Intelligence/book-series/ACCICRC

# Artificial Intelligence, Geographic Information Systems, and Multi-Criteria Decision-Making for Improving Sustainable Development

Edited by

Sujoy Kumar Jana, Kamalakanta Muduli, Indrajit Pal, and Purushottam Meena

CRC Press
Taylor & Francis Group
Boca Raton London New York

CRC Press is an imprint of the
Taylor & Francis Group, an **informa** business
AN AUERBACH BOOK

Designed cover image: Web Large Image (Public); shutterstock

First edition published 2026
2385 NW Executive Center Drive, Suite 320, Boca Raton FL 33431

and by CRC Press
4 Park Square, Milton Park, Abingdon, Oxon, OX14 4RN

*CRC Press is an imprint of Taylor & Francis Group, LLC*



ISBN: 9781032669977 (hbk)
ISBN: 9781032683935 (pbk)
ISBN: 9781032683928 (ebk)

DOI: 10.1201/9781032683928

Typeset in Times
by KnowledgeWorks Global Ltd.

# Contents

# Foreword

The book entitled *Artificial Intelligence, Geographic Information Systems, and Multi-Criteria Decision-Making for Improving Sustainable Development* is a pioneering piece of work. The book is informative and discusses the current discourse on decision-making encompassing contemporary trends of study, research, and applications. The unique work introduces the applications of artificial intelligence (AI) and multi-criteria decision-making (MCDM) to geographic information systems (GIS) to help achieve improvements in sustainable development of teaching, research, and innovation. The book reflects the scientific outputs and their applications in a lucid manner. The chapters reflect a chronological breadth of technological advancement, especially the emergence of the AI revolution in GIS with major implications for teaching, research, and applications in a non-technical and witty style.

The combination of GIS and MCDM generates an excellent analysis tool that results in the creation of an extensive cartographic and alphanumeric database that will later be used to simplify solutions to problems using multi-criteria methodologies.

The authors provide an easy-to-understand basic introduction to AI relevant to GIS in various dimensions. The application of AI in this book is very unique and reflects the integration with GIS and MCDM, which creates a reader-friendly perspective. The book will be helpful for advancing sustainable development in modern infrastructures, including transportation, finance, urban and rural planning, health care, and education.

I take this opportunity to thank and admire the authors and editors for their hard work in this exemplary work, which will serve as a reference book for the higher education community, practitioners, and decision-makers.

**Professor Ora Renagi OL**
*Vice Chancellor*
*Papua New Guinea University of Technology*

# Foreword

# Preface

The Asia-Pacific region, home to some of the world's fastest growing economies, faces a range of complex challenges, including environmental degradation, the increasing frequency of natural hazards, and rapid urbanization. Addressing these issues requires innovative, interdisciplinary approaches to promote sustainable development and enhance resilience. Geographic information systems (GIS), when combined with multi-criteria decision-making (MCDM) techniques and advanced technologies such as artificial intelligence (AI), offer powerful tools to tackle these multifaceted problems. AIGIS integrates AI with GIS to derive insights from geospatial data. AI, including machine learning and deep learning, enables systems to simulate human intelligence, adapt, and learn from data. GIS, on the other hand, manages and analyses geographically referenced information. By combining AI techniques with GIS, this fusion enhances data analysis, visualization, and decision-making. Despite growing research in the convergence of AI and GIS, a unified framework for both research and applied objectives remains underdeveloped. This book provides a comprehensive exploration of how these integrated tools can support decision-making processes aimed at advancing sustainable development in the Asia-Pacific region, drawing on research and insights from diverse disciplines.

The aim of this volume is to offer a comprehensive overview of how GIS, MCDM, and AI can be employed for resilience building and sustainable development in key areas such as disaster risk reduction, environmental monitoring, urban planning, and natural resource management. Through diverse case studies and theoretical explorations, this book highlighted the value of integrated geospatial tools in facilitating informed decision-making and fostering resilient societies in the face of evolving challenges. The chapters in this volume cover a wide range of topics, from site-soil-geology assessments in Fiji to flood risk analysis in Hong Kong, from air quality management in Delhi during the COVID-19 pandemic to vegetation health monitoring in Thailand. Each chapter presents innovative approaches to using GIS and MCDM to address specific problems, showcasing both the diversity of challenges in the region and the versatility of these tools. Other chapters explore the application of these tools in solid waste management, hydropower site selection, structural health monitoring, and environmental governance.

This edited volume brings together the work of academicians, practitioners, and decision-makers from across the region. Each contribution reflects the growing recognition towards effective and sustainable solutions to complex problems, which require a multidimensional approach, integrating scientific, economic, and social considerations. This book aims to equip decision-makers with the tools and frameworks needed to make more resilient and sustainable, considering the interconnectedness of multifaceted challenges. This volume will serve as a valuable resource for researchers, practitioners, and policymakers in fields such as geoinformatics, environmental science, urban planning, disaster management, and operations management. By providing the latest research and practical applications of MCDM, AI, and

GIS, the volume is targeted to contribute to ongoing efforts to build a more sustainable and resilient future for the Asia-Pacific region.

**Sujoy Kumar Jana**
*Papua New Guinea University of Technology, Papua New Guinea*

**Kamalakanta Muduli**
*Papua New Guinea University of Technology, Papua New Guinea*

**Indrajit Pal**
*Asian Institute of Technology, Thailand*

**Purushottam Meena**
*School of Business, College of Charleston, Charleston South Carolina, USA*

# Editors

**Sujoy Kumar Jana** is professor of the GIS section, School of Surveying and Land Studies at the Papua New Guinea University of Technology.

**Kamalakanta Muduli** is associate professor in the School of Mechanical Engineering, Papua New Guinea University of Technology.

**Indrajit Pal** is associate professor and chair of the Disaster Preparedness, Mitigation and Management and Deputy Director of the Research Center "South- and Southeast Asia Multi-disciplinary Applied Research Network on Transforming Societies of Global South (SMARTS)" at the Asian Institute of Technology, Thailand.

**Purushottam Meena** is associate professor of the Supply Chain Management Faculty in the School of Business at the College of Charleston, South Carolina, USA.

# Contributors

**Bashir Ahmed**
Jahangirnagar University
Dhaka, Bangladesh

**Rituparna Acharyya**
Kazimierz Wielki University
Bydgoszcz, Poland

**Rituparna Banerjee**
University of Burdwan
Burdwan, India

**Sunanda Batabyal**
Burdwan University, Bardhaman
West Bengal, India

**Ganni S. V. S. Aditya Bharadwaz**
Asian Institute of Technology
Khlong Nueng, Thailand

**Gouri Sankar Bhunia**
Seacom Skills University
West Bengal, India

**Nilanjan Das**
Burdwan University, Bardhaman
West Bengal, India

**William Feresi**
University of the South Pacific, Laucala Campus
Suva, Fiji Islands

**Sohini Gangopadhyay**
University of Burdwan
Burdwan, India

**Suvojit Ganguly**
Vellore Institute of Technology (VIT) University
Vellore, India

**Subhajit Ghosh**
Asian Institute of Technology
Khlong Nueng, Thailand

**Jyoti Prakash Hati**
Kazimierz Wielki University
Bydgoszcz, Poland

**Zaw Min Htet**
Asian Institute of Technology
Khlong Nueng, Thailand

**Sheikh Tawhidul Islam**
Beihang University
Hangzhou, China

**Sujoy Kumar Jana**
The Papua New Guinea University of Technology
Lae, Papua New Guinea

**Nandanavanam V. B. S. Sudarsan Karthikeya**
Asian Institute of Technology
Khlong Nueng, Thailand

**Nosare Maika**
PNG University of Technology
Lae, Papua New Guinea

**Sudipto Mandal**
University of Burdwan
Burdwan, India

**Sudipta Mukherjee**
Dr. Viswanath Karad MIT World Peace University
Punjab, India

**Anirban Mukhopadhyay**
Kazimierz Wielki University
Bydgoszcz, Poland

**Indrajit Pal**
Asian Institute of Technology
Khlong Nueng, Thailand

**Subodh Chandra Pal**
University of Burdwan
Burdwan, India

**Manju Pandey**
Government College Barpali, Korba
Chhattisgarh, India

**Alak Kumar Patra**
PNG University of Technology
Lae, Papua New Guinea

**Niloy Pramanick**
Asian Institute of Technology
Khlong Nueng, Thailand

**Malay Pramanik**
Asian Institute of Technology
Khlong Nueng, Thailand

**Subhankar Roy**
Jalangi B. Ed College
West Bengal, India

**Tingneyuc Sekac**
PNG University of Technology
Lae, Papua New Guinea

**Saroj Senapati**
Atal Bihari Vajpayee Viswavidyalaya
Chhattisgarh, India

**Arun Kumar Singh**
PNG University of Technology
Lae, Papua New Guinea

**Malini Singh**
Amity University
Noida, India

**Tushar Ashok Tingote**
Dr. Babasaheb Ambedkar Marathwada University
Maharashtra, India

**Adithya Valivety**
Geospatial Data Engineer at Asian Institute of Technology Bangkok
Bangkok, Thailand

**Bhanu Harsha Vardhan Nandamuri**
Department of Computer Science from the Asian Institute of Technology Bangkok
Bangkok, Thailand

**Joeli Varo**
University of the South Pacific, Laucala Campus
Suva, Fiji Islands

**Doni Wainoqolo**
University of the South Pacific, Laucala Campus
Suva, Fiji Islands

# 1 Pathways for Sustainable Development and Multi-Criteria Decision-Making Using AI and GIS

*Arun Kumar Singh and Alak Kumar Patra*

## 1.1 INTRODUCTION

The quest for sustainable development is characterized by the need to harmonize economic growth, social equity, and environmental protection. In this context, decision-makers are often confronted with complex, multifaceted problems that require comprehensive evaluation and strategic planning. Multi-criteria decision-making (MCDM) frameworks have emerged as essential tools for navigating these complexities, allowing stakeholders to evaluate and prioritize multiple alternatives based on a set of diverse and often conflicting criteria.

Artificial Intelligence (AI) and Geographic Information Systems (GIS) have revolutionized the capabilities of MCDM in sustainable development. AI, with its advanced computational techniques such as machine learning, neural networks, and predictive analytics, can process and analyze large volumes of data, identify patterns, and generate forecasts with high accuracy. These capabilities are crucial for understanding and predicting the impacts of various development scenarios, optimizing resource allocation, and enhancing the efficiency of decision-making processes. GIS, on the other hand, offers powerful tools for spatial analysis and visualization. By providing a geographic context, GIS enables the examination of spatial patterns and relationships, which is critical for tasks such as land use planning, environmental impact assessment, and disaster risk management. The ability to visualize complex data spatially allows for a more intuitive and comprehensive understanding of the geographic implications of different development pathways. The integration of AI and GIS in MCDM frameworks facilitates the development of sustainable development pathways that are data-driven and evidence-based. This integrated approach supports the identification, evaluation, and implementation of strategies that balance economic, social, and environmental objectives. It also enhances the ability to monitor and adapt these strategies in response to changing conditions, ensuring long-term sustainability and resilience (Shao et al. 2020).

This introduction outlines the significance of combining AI and GIS within MCDM frameworks to address the intricate challenges of sustainable development.

DOI: 10.1201/9781032683928-1

By leveraging the strengths of these advanced technologies, policymakers and planners can make more informed, effective, and sustainable decisions, paving the way for a balanced and resilient future. Sustainable development pathways aim to balance economic growth, social inclusion, and environmental sustainability. MCDM methods are used to evaluate and prioritize different options based on multiple, often conflicting criteria. The integration of AI and GIS can enhance the process by providing advanced tools for data analysis, visualization, and decision support (Wang et al. 2021).

### 1.1.1 Key Concepts

Figure 1.1 describes the details about key concepts.

- **Sustainable Development Pathways**
  *Economic Growth*: Enhancing productivity and efficiency while maintaining ecological balance.
  *Social Inclusion*: Ensuring equitable access to resources and opportunities for all societal segments.
  *Environmental Sustainability*: Protecting natural resources and reducing ecological footprints.
- **Multi-Criteria Decision-Making (MCDM)**
  *Goal*: To provide a structured approach for evaluating and comparing multiple alternatives based on a set of criteria.
  *Common Methods*: Analytical Hierarchy Process (AHP), Technique for Order Preference by Similarity to Ideal Solution (TOPSIS), and Multi-Attribute Utility Theory (MAUT).

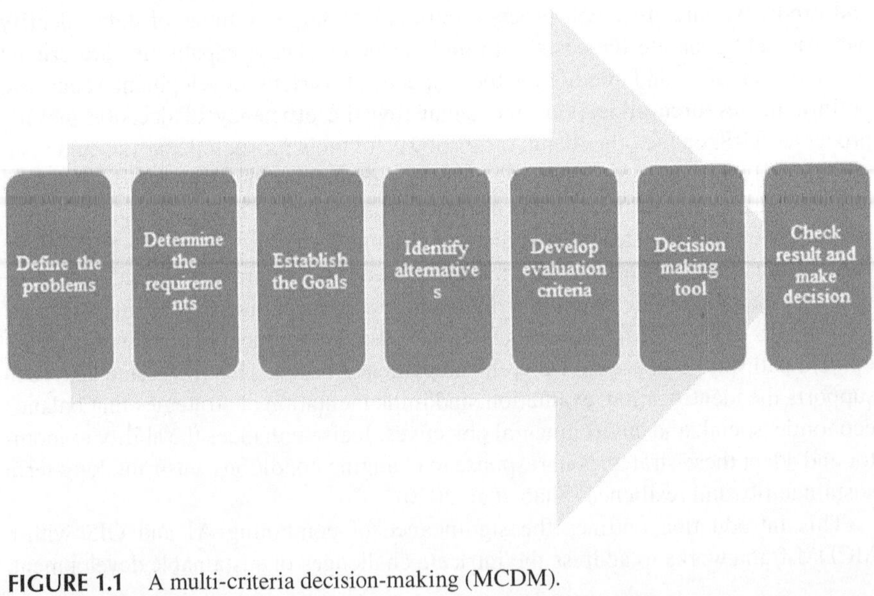


FIGURE 1.1 A multi-criteria decision-making (MCDM).

- **Artificial Intelligence (AI)**
  *Role*: AI can automate and optimize decision-making processes by analyzing large datasets, identifying patterns, and making predictions.
  *Applications*: Machine learning algorithms, neural networks, and expert systems can be employed for predictive modeling, risk assessment, and scenario analysis.
- **Geographic Information Systems (GIS)**
  *Role*: GIS enables the spatial analysis and visualization of geographical data, which is crucial for understanding environmental and socio-economic factors.
  *Applications*: Mapping land use changes, assessing environmental impacts, and planning infrastructure development.

  Applications of AI and GIS in sustainable development.
- **Environmental Monitoring and Management**
  *AI*: Machine learning algorithms can predict pollution levels, identify deforestation patterns, and model climate change impacts.
  *GIS*: Spatial analysis tools can map vulnerable areas, track changes in land use, and assess the distribution of natural resources.

  Figure 1.2 shows the spatial multi-criteria analysis views and other natural resources are given below in detail.
- **Urban Planning and Infrastructure Development**
  *AI*: Predictive models can optimize transportation networks, energy consumption, and waste management systems.

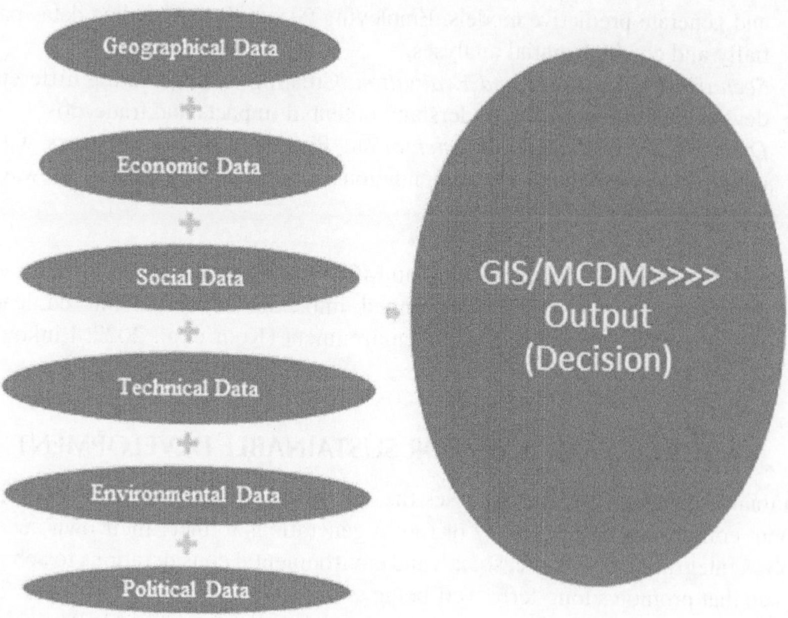


**FIGURE 1.2** Spatial multi-criteria analysis.

*GIS*: Tools can visualize urban growth patterns, evaluate site suitability for new developments, and manage disaster risks.

- **Agriculture and Food Security**
  *AI*: Precision agriculture techniques can enhance crop yield predictions, pest detection, and resource allocation.
  *GIS*: Spatial analysis can assess soil health, monitor crop conditions, and plan irrigation systems.
- **Natural Disaster Management**
  *AI*: Algorithms can predict the occurrence of natural disasters, model their impacts, and optimize emergency response strategies.
  *GIS*: Mapping tools can identify high-risk areas, track the progression of disasters, and coordinate relief efforts.
- **Water Resource Management**
  *AI*: Predictive analytics can forecast water demand, detect leaks, and optimize distribution systems.
  *GIS*: Tools can map water sources, assess water quality, and plan for sustainable usage.

### 1.1.2 Integrating AI and GIS for MCDM in Sustainable Development

- *Data Collection and Integration*: Collecting and integrating diverse datasets from sensors, satellite imagery, surveys, and administrative records.
- *Criteria Definition and Weighting*: Defining relevant criteria for decision-making and assigning weights based on their importance.
- *Modeling and Analysis*: Using AI algorithms to analyze data, identify trends, and generate predictive models. Employing GIS tools to visualize data spatially and conduct spatial analyses.
- *Scenario Development and Evaluation*: Creating and evaluating different development scenarios to understand potential impacts and trade-offs.
- *Decision Support and Implementation*: Providing decision-makers with actionable insights and recommendations. Implementing chosen pathways and monitoring their outcomes.

The strengths of AI and GIS within an MCDM framework, sustainable development efforts can be more effectively planned, implemented, and monitored, leading to better outcomes for society and the environment (Kour et al. 2022; Linkov and Moberg 2011).

## 1.2 BASICS OF PATHWAYS FOR SUSTAINABLE DEVELOPMENT

Sustainable development encompasses the idea of meeting the needs of the present without compromising the ability of future generations to meet their own needs. It involves integrating economic, social, and environmental considerations to achieve a balance that promotes long-term well-being and prosperity for all (Abidi et al. 2022; Kokkinos et al. 2020). Here are some basics of pathways for sustainable development (Table 1.1).

**TABLE 1.1**
**Pathways for Sustainable Development**

| Aspect | Description |
|---|---|
| Triple Bottom Line Approach | Economic, social, and environmental considerations are integrated to achieve balanced development. |
| Principles of Sustainability | Minimize environmental impact, promote social equity, ensure economic viability, respect diversity. |
| Interdisciplinary Approach | Collaboration across disciplines (e.g., environmental science, economics, sociology, technology). |
| Global Goals and Frameworks | United Nations SDGs provide a global roadmap for sustainable development by 2030. |
| Stakeholder Engagement | Involving communities, businesses, governments, NGOs to identify priorities and co-create solutions. |
| Adaptive Management | Continuous learning, monitoring, and adjustment of policies based on new information and feedback. |
| Technology and Innovation | Leveraging tech like renewable energy, sustainable agriculture, digital solutions for sustainable outcomes. |
| Policy and Governance | Effective regulations and incentives to promote sustainability in decision-making and investments. |
| Education and Awareness | Promoting knowledge and capacity-building for sustainable practices and lifestyles. |
| Policy and Governance | Effective regulations and incentives to promote sustainability in decision-making and investments. |

Sustainable development is a holistic approach that aims to meet the needs of the present generation without compromising the ability of future generations to meet their own needs. This abstract explores the foundational principles, strategies, and frameworks essential for navigating the complexities of sustainable development (Figure 1.3).

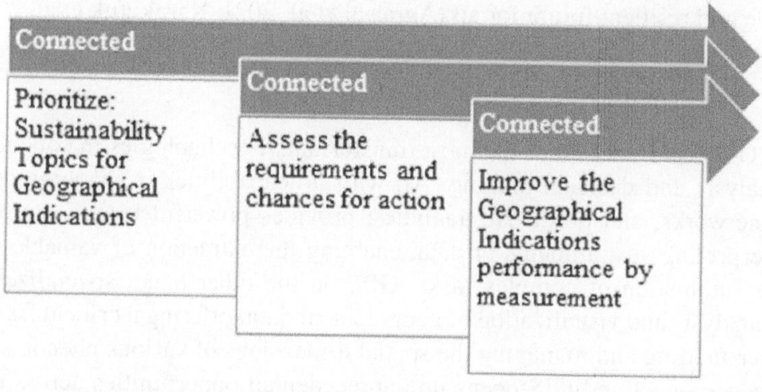


**FIGURE 1.3** Basic components of strategic roadmap for sustainability.

The concept of sustainable development encompasses three interconnected dimensions: economic prosperity, social equity, and environmental stewardship. These dimensions are often referred to as the triple bottom line, emphasizing their interdependence and the need for balanced decision-making that considers all three aspects simultaneously. Key principles guiding pathways for sustainable development include minimizing environmental degradation, promoting social inclusivity, ensuring economic viability, and respecting cultural diversity. Interdisciplinary collaboration among fields such as environmental science, economics, sociology, technology, and policymaking is crucial for addressing multifaceted challenges and designing integrated solutions (Aboul Ella Hassanien and Chiranji 2021; Klör et al. 2018).

The United Nations Sustainable Development Goals (SDGs) provide a global framework for collective action toward sustainable development by 2030. These goals address critical issues such as poverty alleviation, climate action, sustainable cities, biodiversity conservation, and responsible consumption and production. Stakeholder engagement plays a pivotal role in sustainable development, ensuring inclusivity, transparency, and accountability in decision-making processes. Adaptive management strategies enable continuous learning, monitoring, and adjustment of policies based on evolving information and feedback, thereby enhancing resilience and effectiveness. Technological innovation, including advancements in renewable energy, sustainable agriculture, and digital solutions, accelerates progress toward sustainable outcomes. Effective policy and governance frameworks are essential for translating sustainable development principles into actionable plans and regulations that incentivize sustainable practices and investments. Education and awareness initiatives are integral to fostering a culture of sustainability, empowering individuals and communities to adopt responsible behaviors and lifestyles (Kitchenham 2004).

In conclusion, embracing the basics of pathways for sustainable development requires a comprehensive and integrated approach that balances economic, social, and environmental considerations. By incorporating these principles into decision-making processes and everyday practices, stakeholders can contribute to building a more equitable, resilient, and sustainable future for all. By embracing these basics and integrating them into planning, decision-making, and everyday actions, stakeholders can contribute to advancing sustainable development and achieving a more equitable and resilient future for all (Agrawal et al. 2021; Karakutuk et al. 2021).

### 1.2.1 AI and GIS Applications/Techniques

AI and GIS represent two of the most transformative technologies in contemporary data analysis and decision-making. AI, with its capabilities in machine learning, neural networks, and predictive analytics, provides powerful tools for processing and interpreting vast amounts of data, enabling the extraction of valuable insights and the automation of complex tasks. GIS, on the other hand, specializes in the spatial analysis and visualization of geographical data, offering a critical framework for understanding and managing the spatial dimensions of various phenomena. The convergence of AI and GIS opens up unprecedented opportunities across multiple domains, enhancing the capacity to address intricate challenges through informed decision-making. This integration leverages the strengths of both technologies: AI's analytical prowess and GIS's spatial context, thereby providing a comprehensive approach to data-driven problem-solving. By combining AI's predictive and optimization abilities with GIS's spatial analysis, stakeholders can develop more efficient, effective, and sustainable solutions (Alimohammadlou and Khoshsepehr 2023; Aloui et al. 2021; Kamari et al. 2020).

Applications of AI and GIS span diverse fields, each benefiting from the enhanced analytical capabilities and spatial insights these technologies provide. In environmental management, they aid in predicting and mitigating the impacts of climate change and pollution. In urban planning, they optimize infrastructure development and smart city initiatives. In agriculture, they enhance precision farming techniques and food security measures. In disaster management, they improve preparedness and response strategies. Finally, in water resource management, they optimize usage and conservation practices (Anderson et al. 2018; Arana-Catania et al. 2021; Janssen and van Herwijnen 2020).

### 1.2.2 AI Techniques

1. **Machine Learning**
   *Description*: Machine learning algorithms enable systems to learn from data and make predictions or decisions without explicit programing.
   *Applications*: Predictive modeling, classification, clustering, anomaly detection, and recommendation systems.
   *Examples*: Support Vector Machines (SVM), Random Forest, Neural Networks, and Deep Learning.
2. **Natural Language Processing (NLP)**
   *Description*: NLP involves the interaction between computers and humans using natural language, enabling machines to process, analyze, and understand human language.
   *Applications*: Sentiment analysis, language translation, chatbots, and text summarization.
   *Examples*: Named Entity Recognition (NER), Text Classification, Language Generation models like GPT-3.
3. **Computer Vision**
   *Description*: Computer vision enables machines to interpret and understand visual information from the world.
   *Applications*: Object detection, image classification, facial recognition, medical image analysis, and autonomous vehicles.
   *Examples*: Convolutional Neural Networks (CNN), Image Segmentation, and Optical Character Recognition (OCR).
4. **Reinforcement Learning**
   *Description*: Reinforcement learning involves training algorithms to make sequential decisions by interacting with an environment and learning from feedback.
   *Applications*: Game playing (e.g., AlphaGo), robotics, autonomous systems, and resource management.

*Examples*: Q-Learning, Deep Q-Networks (DQN), and Policy Gradient methods.

5. **Predictive Analytics**

*Description*: Predictive analytics uses statistical techniques and machine learning to analyze current and historical data to make predictions about future events.

*Applications*: Forecasting, risk assessment, demand forecasting, and predictive maintenance.

*Examples*: Time series forecasting models like ARIMA, and Prophet.

#### 1.2.2.1 GIS Techniques

1. **Spatial Analysis**

*Description*: Spatial analysis examines geographic data to understand patterns, relationships, and trends.

*Applications*: Buffering, overlay analysis, interpolation, and proximity analysis.

*Examples*: Spatial statistics (e.g., Moran's I), Hotspot analysis, and Spatial regression.

2. **Geospatial Data Modeling**

*Description*: Geospatial data modeling involves representing real-world features, phenomena, and relationships in a digital environment.

*Applications*: 3D modeling, network analysis, land use planning, and hydrological modeling.

*Examples*: Network Analysis (e.g., shortest path), Digital Elevation Models (DEMs), and Hydrological Models (e.g., SWAT).

3. **Remote Sensing**

*Description*: Remote sensing involves collecting data about Earth's surface without direct physical contact, typically from satellites or aircraft.

*Applications*: Land cover classification, vegetation monitoring, urban growth analysis, and disaster assessment.

*Examples*: Image classification (e.g., supervised and unsupervised), Change detection, and Thermal imaging.

4. **Cartography and Visualization**

*Description*: Cartography involves the creation and design of maps, while visualization uses graphical techniques to represent geospatial data.

*Applications*: Map design, thematic mapping, spatial storytelling, and interactive visualization.

*Examples*: Heatmaps, Choropleth maps, and Story maps.

5. **Spatial Decision Support Systems (SDSS)**

*Description*: SDSS integrates GIS with decision-making processes to support complex spatial decisions.

*Applications*: Site selection, environmental impact assessment, disaster management, and urban planning.

*Examples*: Multi-Criteria Decision Analysis (MCDA), Location-Allocation models, and Scenario planning.

#### 1.2.2.1.1 Integration of AI and GIS Techniques

The integration of AI and GIS techniques enhances decision-making processes by combining AI's advanced analytical capabilities with GIS's spatial context. This integration enables more comprehensive analyses, predictive modeling with spatial components, and automation of complex spatial tasks. By leveraging both AI and GIS techniques, organizations can achieve more accurate insights, optimize resource management, and develop sustainable solutions across various domains. This introduction highlights the significant role of AI and GIS applications in transforming decision-making processes across various sectors, demonstrating how their integration facilitates more informed, efficient, and sustainable outcomes (Halsband 2022).

## 1.3 AI AND GIS APPLICATIONS WITH PATHWAYS FOR SUSTAINABLE DEVELOPMENT AND MCDM

Sustainable development involves meeting the needs of the present without compromising the ability of future generations to meet their own needs. MCDM frameworks can be valuable in navigating the complexities of sustainable development by considering various factors and trade-offs (Ascher et al. 2022; Eitel-Porter 2020; Goel et al. 2020). Here are some pathways and considerations for integrating sustainable development and MCDM:

1. *Identifying Stakeholders and Objectives*: Begin by identifying all stakeholders involved in the decision-making process, including communities, businesses, governments, and NGOs. Define clear objectives that align with SDGs, such as environmental conservation, social equity, economic prosperity, and technological advancement.
2. *Criteria Development*: Develop a set of criteria that reflect the different dimensions of sustainability relevant to the decision at hand. These criteria may include environmental impact, social inclusivity, economic viability, technological feasibility, cultural preservation, and policy compliance. Each criterion should be measurable and relevant to the stakeholders' concerns.
3. *Gathering Data and Information*: Collect relevant data and information to assess the performance of different alternatives against the established criteria. This may involve conducting environmental assessments, socioeconomic analyses, feasibility studies, and risk assessments. Data-driven decision-making ensures that choices are based on evidence and comprehensive analysis.
4. *Multi-Criteria Evaluation Methods*: Utilize appropriate MCDM methods to evaluate and compare alternative courses of action. Common techniques include AHP, Analytic Network Process (ANP), MAUT, and outranking methods like ELECTRE and PROMETHEE. These methods help in quantifying qualitative factors and prioritizing alternatives based on their overall performance across multiple criteria.

5. *Scenario Analysis and Sensitivity Testing*: Conduct scenario analysis to evaluate how different future scenarios may impact the outcomes of sustainable development initiatives. Sensitivity testing helps in understanding the robustness of decisions under varying conditions and uncertainties, ensuring resilience and adaptability in the face of change.
6. *Stakeholder Engagement and Transparency*: Engage stakeholders throughout the decision-making process to ensure inclusivity, transparency, and accountability. Stakeholder inputs provide valuable insights, enhance the legitimacy of decisions, and foster collective ownership of sustainable development initiatives.
7. *Monitoring and Adaptation*: Implement mechanisms for monitoring the implementation and outcomes of chosen strategies over time. Regular evaluation allows for adaptive management, where adjustments can be made based on feedback, new information, and changing circumstances to enhance effectiveness and achieve long-term sustainability goals.
8. *Policy Integration and Institutional Capacity*: Integrate sustainable development considerations into policy frameworks and strengthen institutional capacity for effective implementation. This involves aligning regulatory frameworks, promoting innovation and knowledge sharing, and building partnerships across sectors and jurisdictions.

By integrating MCDM with sustainable development principles, stakeholders can make informed choices that balance environmental, social, and economic considerations, thereby promoting holistic and resilient pathways to sustainable development.

The integration of AI and GIS is revolutionizing pathways for sustainable development and MCDM. Sustainable development aims to balance economic growth, social inclusion, and environmental protection, necessitating complex decision-making processes that can account for multiple, often conflicting criteria. AI and GIS provide the necessary tools and methodologies to enhance these processes by offering advanced data analysis, visualization, and decision support capabilities (Ashley et al. 2003).

AI brings to the table powerful computational techniques such as machine learning, neural networks, and predictive analytics. These capabilities enable the processing and analysis of vast and diverse datasets, uncovering patterns, predicting future trends, and optimizing resource allocation. GIS complements this by offering robust spatial analysis and visualization tools, essential for understanding geographic patterns and relationships. Together, AI and GIS facilitate the development of more effective and sustainable strategies by integrating spatial and non-spatial data, optimizing land use, monitoring environmental changes, and supporting disaster risk management (Astobiza et al. 2021; Dube et al. 2021).

The integration of AI and GIS within MCDM frameworks allows for a comprehensive evaluation of different development scenarios, considering multiple criteria such as economic viability, social equity, and environmental impact (Table 1.2). This approach supports policymakers and planners in making data-driven, evidence-based decisions that are critical for achieving long-term sustainability goals (Bokhari and Myeong 2022; Bregaglio et al. 2022; Cervelli et al. 2020).

**TABLE 1.2**
**AI and GIS Applications in Sustainable Development and MCDM**

| Domain | AI Applications | GIS Applications |
|---|---|---|
| Environmental Management | • Predictive Modeling: AI algorithms predict pollution levels, climate change impacts, and deforestation patterns.<br>• Pattern Recognition: Machine learning models identify and monitor changes in land use and vegetation cover.<br>• Optimization: AI helps optimize resource allocation for conservation efforts and pollution control. | • Spatial Analysis: GIS tools map and analyze environmental data, such as habitat distribution and ecosystem health.<br>• Impact Assessment: GIS visualizes potential impacts of environmental policies and projects on different regions.<br>• Resource Management: GIS supports the planning and management of natural resources like water, forests, and wildlife. |
| Urban Planning and Infrastructure Development | • Predictive Analytics: AI models forecast urban growth, traffic patterns, and infrastructure needs.<br>• Smart Cities: AI-driven systems optimize energy usage, waste management, and public transportation in urban areas.<br>• Risk Analysis: AI assesses risks related to infrastructure projects, including financial, environmental, and social factors. | • Site Suitability Analysis: GIS identifies optimal locations for new developments based on various criteria.<br>• Zoning and Land Use Planning: GIS helps in creating and managing zoning plans and land use maps.<br>• Disaster Risk Management: GIS maps high-risk areas and supports disaster preparedness and response planning. |
| Agriculture and Food Security | • Precision Agriculture: AI analyzes data from sensors and drones to optimize irrigation, fertilization, and pest control.<br>• Yield Prediction: Machine learning models predict crop yields based on weather, soil, and crop data.<br>• Supply Chain Optimization: AI improves the efficiency of agricultural supply chains, from production to distribution. | • Soil and Crop Mapping: GIS maps soil types, crop conditions, and productivity zones for better agricultural planning.<br>• Land Use and Crop Rotation: GIS supports planning for optimal land use and crop rotation schedules.<br>• Water Resource Management: GIS helps plan and manage irrigation systems and water resource distribution. |
| Natural Disaster Management | • Disaster Prediction: AI models predict natural disasters like earthquakes, floods, and hurricanes.<br>• Impact Simulation: AI simulates disaster scenarios to plan effective response strategies. | • Risk Mapping: GIS maps areas prone to natural disasters, aiding in risk assessment and mitigation planning.<br>• Emergency Response Coordination: GIS supports the coordination of emergency response efforts during disasters. |

*(Continued)*

**TABLE 1.2 (Continued)**
**AI and GIS Applications in Sustainable Development and MCDM**

| Domain | AI Applications | GIS Applications |
|---|---|---|
| Water Resource Management | • Demand Forecasting: AI predicts water demand and usage patterns for efficient water resource management.<br>• Leak Detection: AI detects leaks in water distribution systems to minimize water loss.<br>• Optimization: AI optimizes the distribution of water resources to balance supply and demand. | • Source Mapping: GIS maps water sources, including rivers, lakes, and groundwater reserves.<br>• Quality Monitoring: GIS monitors water quality across different regions, identifying areas of concern.<br>• Sustainable Usage Planning: GIS supports the planning of sustainable water usage practices. |

## 1.4 ADVANTAGES OF AI AND GIS APPLICATIONS IN PATHWAYS FOR SUSTAINABLE DEVELOPMENT AND MCDM

The integration of AI and GIS with pathways for sustainable development and MCDM offers numerous advantages (Bianchini et al. 2019; Chen et al. 2021; de Oliveira Musse et al. 2018). These technologies can significantly enhance the efficiency, accuracy, and effectiveness of decision-making processes, leading to better sustainable development outcomes (Tables 1.3 and 1.4).

**TABLE 1.3**
**AI and GIS Applications in Pathways for Sustainable Development**

| Aspect | AI Applications | GIS Applications | Combined Advantages |
|---|---|---|---|
| Data Processing and Analysis | • Speed and Efficiency: Rapid processing of large datasets. | • Comprehensive Datasets: Integrates diverse datasets for holistic analysis. | • Multi-Criteria Evaluation: Combines spatial data and predictive analytics for comprehensive assessments. |
| | • Pattern Recognition: Identifies complex patterns and correlations. | • Interoperability: Works with various data formats and sources. | • Stakeholder Engagement: Enhances understanding and consensus-building through visualizations and predictive models. |
| Predictive Modeling | • Forecasting: Creates predictive models for future scenario planning. | • Visualization: Provides powerful tools for visualizing spatial data. | • Real-Time Analysis: Processes real-time data for up-to-date decision-making and timely interventions. |
| | • Scenario Analysis: Simulates different scenarios for evaluating potential outcomes. | • Mapping: Creates detailed maps highlighting critical areas for sustainable development. | • Adaptive Management: Supports continuous refinement of policies based on emerging data and feedback. |

*(Continued)*

**TABLE 1.3 (Continued)**
**AI and GIS Applications in Pathways for Sustainable Development**

| Aspect | AI Applications | GIS Applications | Combined Advantages |
|---|---|---|---|
| Optimization | • Resource Allocation: Optimizes the use of resources for achieving sustainability goals.<br>• Optimization | • Land Use Planning: Aids in effective land use planning and preserving natural habitats.<br>• Resource Allocation: Optimizes the use of resources for achieving sustainability goals. | • Risk Assessment and Mitigation: Identifies vulnerable areas and assesses potential impacts for targeted interventions.<br>• Land Use Planning: Aids in effective land use planning and preserving natural habitats. |
| Automation | • Routine Tasks: Automates repetitive tasks to free up human resources for complex tasks. | | |
| Data Integration | | • Integration: Integrates various datasets into a single platform for a comprehensive view. | |
| Monitoring and Management | | • Monitoring: Monitors environmental changes, urbanization, and climate impacts. | |
| Policy and Governance | • Decision Support: AI enhances decision support systems for effective policymaking. | • Spatial Governance: Supports the development of spatially informed policies. | |
| Education and Awareness | • Predictive Insights: AI provides predictive insights for educational purposes. | • Geospatial Awareness: GIS enhances geospatial understanding and awareness. | |

**TABLE 1.4**
**Disadvantages of AI and GIS Applications in Pathways for Sustainable Development and Multi-Criteria Decision-Making (MCDM)**

| Aspect | AI Applications | GIS Applications | Combined Disadvantages |
|---|---|---|---|
| Data Quality and Availability | • Data Dependency: AI models require high-quality, extensive datasets, which may not always be available or accurate. | • Data Accuracy: GIS analyses depend on the accuracy and resolution of spatial data, which can be limited or outdated. | • Data Integration: Combining datasets from different sources can be challenging due to inconsistencies and varying data standards. |

*(Continued)*

**TABLE 1.4 (Continued)**
**Disadvantages of AI and GIS Applications in Pathways for Sustainable Development and Multi-Criteria Decision-Making (MCDM)**

| Aspect | AI Applications | GIS Applications | Combined Disadvantages |
|---|---|---|---|
| Technical Complexity | • Complex Algorithms: AI involves complex algorithms that may be difficult for non-experts to understand and implement. | • Technical Expertise: GIS requires specialized knowledge for effective use, which can be a barrier for some stakeholders. | • Interdisciplinary Challenges: Integrating AI and GIS requires collaboration across different technical fields, which can be complex to manage. |
| Cost and Resources | • High Costs: Implementing AI solutions can be expensive, requiring significant investment in hardware, software, and expertise. | • Infrastructure Costs: GIS systems can be costly to set up and maintain, particularly in resource-limited settings. | • Resource Intensity: Both technologies can be resource-intensive, requiring substantial financial and human resources for effective deployment. |
| Ethical and Privacy Concerns | • Bias and Fairness: AI algorithms can perpetuate biases present in the training data, leading to unfair or biased outcomes. | • Privacy Issues: GIS data collection and usage can raise privacy concerns, especially when dealing with sensitive location information. | • Data Security: Ensuring the security and privacy of integrated datasets is a significant challenge. |
| Scalability and Adaptability | • Scalability: AI models may not scale well across different regions or contexts due to variations in data and conditions. | • Context-Specific: GIS solutions may need significant customization for different geographical or socio-economic contexts. | • Integration Challenges: Scaling integrated AI and GIS solutions across diverse contexts can be complex and require significant adaptation. |
| Dependence on Technology | • Over-Reliance: Excessive reliance on AI can reduce human oversight and critical thinking, potentially leading to unforeseen consequences. | • Technical Failures: GIS systems can be prone to technical issues or failures, affecting the reliability of the analysis. | • System Interdependencies: Dependence on integrated systems increases vulnerability to technical failures or disruptions in either component. |

*(Continued)*

**TABLE 1.4 (Continued)**
**Disadvantages of AI and GIS Applications in Pathways for Sustainable Development and Multi-Criteria Decision-Making (MCDM)**

| Aspect | AI Applications | GIS Applications | Combined Disadvantages |
|---|---|---|---|
| Implementation Barriers | • Resistance to Change: Organizations may resist adopting AI due to perceived complexity or fear of job displacement. | • Adoption Challenges: GIS adoption can be hindered by a lack of technical skills or resistance to change among stakeholders. | • Interoperability Issues: Ensuring interoperability between AI and GIS systems can be challenging, especially when integrating legacy systems. |
| Transparency and Explainability | • Black Box Problem: AI algorithms can be opaque, making it difficult to understand how decisions are made, which can undermine trust. | • Complex Outputs: GIS outputs can be complex and difficult for non-specialists to interpret. | • Interpretability: Combining AI and GIS can result in complex systems that are challenging to explain and understand comprehensively. |

## 1.5 CONCLUSION

In conclusion, the integration of AI and GIS with MCDM frameworks presents a promising pathway toward advancing SDGs. AI algorithms enable robust data analysis, predictive modeling, and pattern recognition, thereby enhancing the accuracy and efficiency of decision-making processes. GIS technologies provide spatial data visualization, mapping, and spatial analysis capabilities, which are essential for understanding geographical contexts and optimizing resource allocation. Together, AI and GIS empower stakeholders to evaluate complex environmental, social, and economic factors simultaneously. AI algorithms can analyze vast datasets to identify patterns, correlations, and trade-offs among various criteria, guiding decision-makers toward informed and evidence-based choices. GIS platforms facilitate the visualization of spatial relationships and the integration of geospatial data, enabling stakeholders to assess the spatial impacts of development initiatives and identify optimal locations for sustainable interventions. The combination of AI and GIS fosters scenario analysis and predictive modeling, allowing stakeholders to simulate different development scenarios and anticipate their potential impacts on ecosystems, communities, and economies. This proactive approach supports adaptive management strategies, where decisions can be adjusted in response to changing environmental conditions and stakeholder feedback.

Incorporating AI and GIS into MCDM frameworks also promotes transparency, stakeholder engagement, and interdisciplinary collaboration. By providing accessible and comprehensible insights, these technologies facilitate inclusive decision-making processes that account for diverse perspectives and enhance the accountability of sustainable development initiatives. In core, leveraging AI and GIS in MCDM frameworks enhances the capacity of stakeholders to navigate the complexities of sustainable development effectively. By harnessing the power of data-driven insights and spatial analysis, decision-makers can pursue pathways that promote environmental stewardship, social equity, and economic resilience, thereby fostering sustainable development for present and future generations.

## REFERENCES

Abidi, M. H., Mohammed, M. K., & Alkhalefah, H. (2022). Predictive maintenance planning for Industry 4.0 using machine learning for sustainable manufacturing. Sustainability, 14(6), 3387. DOI: 10.3390/su14063387.

Aboul Ella Hassanien, N. D., & Chiranji, L. C. (2021). "Sustainable Development Goals: Leveraging the Role of Artificial Intelligence and GIS", In: Artificial Intelligence and Machine Learning for Sustainable Development Goals, Springer.

Agrawal, R., Wankhede, V. A., Kumar, A., Luthra, S., & Huisingh, D. (2021). Progress and trends in integrating Industry 4.0 within circular economy: A comprehensive literature review and future research propositions. Business Strategy and the Environment, 31, 559–579.

Alimohammadlou, M., & Khoshsepehr, Z. (2023). The role of Society 5.0 in achieving stainable development: A spherical fuzzy set approach. Environmental Science and Pollution Research, 30(16), 47630–47654.

Aloui, A., Hamani, N., & Delahoche, L. (2021). An integrated optimization approach using a collaborative strategy for sustainable cities freight transportation: A case study. Sustainable Cities and Society, 75, 103331.

Anderson, R., Keshwani, D., Guru, A., Yang, H., Irmak, S., & Subbiah, J. (2018). An integrated modeling framework for crop and biofuel systems using the DSSAT and GREET models. Environmental Modelling & Software, 108, 40–50.

Arana-Catania, M., Lier, F.-A. V., Procter, R., Tkachenko, N., He, Y., Zubiaga, A., & Liakata, M. (2021). Citizen participation and machine learning for a better democracy. Digital Government: Research and Practice, 2(3), 1–22.

Ascher, S., Sloan, W., Watson, I., & You, S. (2022). A comprehensive artificial neural network model for gasification process prediction. Applied Energy, 320, 119289.

Ashley, R. M., Blackwood, D. J., Butler, D., Davies, J., Jowitt, P. W., & Smith, H. (2003). Sustainable decision making for the UK water industry. Proceedings of the Institution of Civil Engineers Engineering Sustainability, Issue ES1, 41–49.

Astobiza, A. M., Toboso, M., Aparicio, M., & López, D. (2021). AI ethics for sustainable development goals. IEEE Technology and Society Magazine, 40, 66–71.

Bianchini, A., Rossi, J., & Pellegrini, M. (2019). Overcoming the main barriers of circular economy implementation through a new visualization tool for circular business models. Sustainability, 11, 6614.

Bokhari, S. A. A., & Myeong, S. (2022). Use of Artificial Intelligence in Smart Cities for Smart Decision-Making: A Social Innovation Perspective. Sustainability, 14, 620. DOI: 10.3390/su14020620.

Bregaglio, S., Savian, F., Raparelli, E., Morelli, D., Epifani, R., Pietrangeli, F., … Manici, L. M. (2022). A public decision support system for the assessment of plant disease infection risk shared by Italian regions. Journal of Environmental Management, 317, 115365.

Cervelli, E., di Perta, E. S., & Pindozzi, S. (2020). Energy crops in marginal areas: Scenario-based assessment through ecosystem services, as support to sustainable development. Ecological Indicators, 113, 106180.

Chen, J., Lim, C. P., Tan, K. H., Govindan, K., & Kumar, A. (2021). Artificial intelligence-based human-centric decision support framework: An application to predictive maintenance in asset management under pandemic environments. Annals of Operations Research, 1–24. https://pubmed.ncbi.nlm.nih.gov/34785834/

de Oliveira Musse, J., Homrich, A. S., de Mello, R., & Carvalho, M. M. (2018). Applying back casting and system dynamics towards sustainable development: The housing planning case for low-income citizens in Brazil. Journal of Cleaner Production, 193, 97–114.

Dube, F., Nzimande, N., & Muzindutsi, P.-F. (2021). Application of artificial neural networks in predicting financial distress in the JSE financial services and manufacturing companies. Journal of Sustainable Finance & Investment, 13(1), 723–743.

Eitel-Porter, R. (2020). Beyond the promise: Implementing ethical AI. AI and Ethics, 1, 73–80.

Goel, P., Jain, P., Pasman, H. J., Pistikopoulos, E. N., & Datta, A. (2020). Integration of data analytics with cloud services for safer process systems, application examples and implementation challenges. Journal of Loss Prevention in The Process Industries, 68, 104316.

Halsband, A. (2022). Sustainable AI and intergenerational justice. Sustainability, 14(7), 1–11.

Janssen, R., & van Herwijnen, M. (2020). Decision support systems for sustainable development: A review and analysis of challenges. Journal: Environmental Modelling & Software. DOI: 10.1016/j.envsoft.2020.104897

Kamari, M. L., Isvand, H., & Nazari, M. A. (2020). Applications of multi-criteria decision-making (MCDM) methods in renewable energy development: A review. Renewable Energy Research and Applications, 1(1), 47–54.

Karakutuk, S. S., Akpinar, S., & Ornek, M. A. (2021). An application of a circular economy approach to design an energy-efficient heat recovery system. Journal of Cleaner Production, 320, 128851.

Kitchenham, B. (2004). Procedures for performing systematic reviews. Keele, UK, Keele University, 33(2004), 1–26.

Klör, B., Monhof, M., Beverungen, D., & Bräuer, S. (2018). Design and evaluation of a model-driven decision support system for repurposing electric vehicle batteries. European Journal of Information Systems, 27(2), 171–188.

Kokkinos, K., Karayannis, V., & Moustakas, K. (2020). Circular bio-economy via energy transition supported by Fuzzy Cognitive Map modeling towards sustainable low-carbon environment. Science of the Total Environment, 721, 137754.

Kour, R., Castaño, M., Karim, R., Patwardhan, A., Kumar, M., & Granström, R. (2022). A human-centric model for sustainable asset management in railway: A case study. Sustainability, 14(2), 936.

Linkov, I., & Moberg, E. (2011). Multi-criteria decision analysis in environmental sciences: Ten years of applications and trends. Journal: Science of the Total Environment, 409(19), 3578–3594.

Shao, M., Han, Z., Sun, J., Xiao, C., Zhang, S., & Zhao, Y. (2020). A review of multi-criteria decision making applications for renewable energy site selection. Renew Energy, 157, 377–403.

Wang, Y., Tan, P. V., & Chen, J. (2021). AI and big data for sustainable development: A systematic review and future research directions. Journal: Journal of Cleaner Production. DOI: 10.1016/j.jclepro.2020.124682

# 2 Sustainable Development through Geospatial Multi-Criteria Decision-Making (MCDM)

## *A Comprehensive Analysis*

*Anirban Mukhopadhyay, Sujoy Kumar Jana, Jyoti Prakash Hati, Rituparna Acharyya, and Indrajit Pal*

## 2.1 INTRODUCTION

In 1984, the "United Nations" (UN) established an autonomous committee with 22 members from developing and rich countries to formulate long-term environmental policies for the worldwide community. The "World Commission on Environment and Development" (WCED, 1987) released the report "Our Common Future," which is generally recognised as critical in establishing sustainable development globally and shaping a common future. In this report, the term "sustainable development" was coined for the first time, which is defined as development that fulfils current demands without jeopardizing future generations' ability to fulfil their demands or requirements. In this report, the commission examined environmental issues raised by development processes from an economic, social, and political standpoint, rather than just from a scientific aspect (Elliott, 2012). Their ideas were focused on integrating development strategies, environmental regulations, and global partnerships to address interconnected environmental issues and growth prospects. In 2015, the UN member nations agreed on the "2030 Agenda for Sustainable Development," which provides a united roadmap for prosperity and peace for humans and the environment currently and in the foreseeable future. The core policy is based on 17 "Sustainable Development Goals" (SDGs), which signify an urgent call for action from all states in a global collaboration. The SDGs aim to eradicate poverty and other hardships, provide education and health facilities, eliminate inequality, and drive economic growth while tackling climate change and safeguarding the environment.

The SDGs have replaced the previously set "Millennium Development Goals" (MDGs) by the UN and put more emphasis on investigators than ever before to deal with climate change, energy efficiency, health, nutrition, and water supply issues. The effective execution of sustainable development depends on understanding resource management and regional challenges (Tabor and Hutchinson, 1994). For


DOI: 10.1201/9781032683928-2

this purpose, numerous social, economic, and environmental issues must be monitored and modelled globally. Remote sensing (RS) is an efficient tool for addressing ecological and socio-economic challenges at a global scale. It allows for monitoring Earth's systems at different spatial and temporal scales, providing prompt and accurate information. RS is particularly useful for gathering data in remote and inaccessible regions, making it the primary surveillance approach (Avtar et al., 2020). Although numerous methods and techniques exist to assess natural resources and the hazards affecting them, the RS technique has been employed since the 1970s due to its affordable acquisition cost and great utility in data collection, analysis, and administration. The methodologies and approaches based on RS have been applied for a range of applications at various time scales across the past few decades (Jensen, 2009). RS provides a unified perspective of spatial information at the local, regional, and global levels, allowing for faster decision-making and action (Jensen and Cowen, 1999). Thus, over the past 20 years, RS has been strongly applied in the agricultural sector (Khanal et al., 2020; Shanmugapriya et al., 2019); forestry (Boyd and Danson, 2005; Lechner et al., 2020); fisheries (Klemas, 2013; Stuart et al., 2011); environmental solutions (Mahrad et al., 2020; Yuan et al., 2020); urban management (Krishnaveni and Anilkumar, 2020; Yang, 2021); disaster mitigation (Eguchi et al., 2008; Joyce et al., 2009); etc.

Currently, numerous assessments are made based on many criteria acquired from expert groups and determined accordingly based on the weightage assigned. There are not just significant challenges that require several criteria, even certain criteria could have an impact on certain challenges; but, to arrive at an ideal solution, all choices must comply with identical criteria, which leads to effectively informed decision-making (Aruldoss et al., 2013). Multi-criteria decision-making (MCDM) deals with the framework and solutions for decision-making and planning issues encompassing numerous criteria. The fundamental purpose of MCDM is to support decision-makers and stakeholders whenever there are several possibilities for resolving a problem. When there is no ideal solution to these problems, it becomes necessary to capitalize on the decision-maker's propensity to choose amongst alternatives. With the advancement of new technology and modern algorithms, MCDM has been used massively in every aspect of social life in the last two decades. Some prominent applications include construction management (Jato-Espino et al., 2014), sustainable planning of energy consumption (Pohekar and Ramachandran, 2004), management of waste disposal (Achillas et al., 2013), transportation projects (Macharis and Bernardini, 2015), site selection (Yap et al., 2019), urban planning (Afshari et al., 2016), and renewable energy development (Kamari et al., 2020).

Although there are ample publications on the MCDM approaches being used in different sectors, there is a deficit of sustainable development studies using MCDM in the literature. Since the implementation of the Sendai Framework in 2015, there has been a growing need for MCDM applications in various sectors of society. Therefore, this study aims to present a comprehensive overview of the research that has already been published utilizing MCDM and RS. This approach will help to synthesize the knowledge we already have on this topic and hope to chalk out a direction for future research on RS-based MCDM studies on sustainable development.

## 2.2 SUSTAINABILITY INDICATORS

Since the late 1980s, as the impact of the environment on society became more prominent, the performance metric related to sustainability has become the focus of mainstream research. Several studies have used "Cost" as the primary indicator to analyse economic growth (Ma et al., 2018; Remery et al., 2012). Further comprehensive economic performance indicators are provided by project return or revenue, which reveals how much additional revenue, could be generated from the project's effective completion. Several techniques are employed to assess profitability, with net present value (NPV) being the most popular strategy (Larson and Gray, 2014). The payback model calculates the time needed to recover an investment and is another important method for evaluating profitability. The "Return on Investment" (ROI) is a further indicator of economic performance (Ma et al., 2020). This metric makes it possible to compare the prospective revenues of various projects.

In addition to economic considerations, environmental performance has been considered regarding project selection. In earlier pertinent research, several sectors, notably manufacturing (Labuschagne and Brent, 2005) and construction (Dobrovolskienė and Tamošiūnienė, 2016), aimed at incorporating environmental assessment into project management. In general, environmental performance is utilized as a scenario that assists with decision-making within project management (Jugend and Figueiredo, 2017). Some research used project "Life Cycle Assessment" (LCA) and eco-design methodologies in "Project Portfolio Selection" (PPS) (Bovea and Pérez-Belis, 2012; Brones and Monteiro de Carvalho, 2015). "Environmental Quality Function Deployment" (EQFD) (Bovea and Pérez-Belis, 2012), "Materials, Energy, and Toxicity" (MET) matrix (Byggeth and Hochschorner, 2006), and "Environmental Failure Mode Effects Analysis" (E-FMEA) (Knight and Jenkins, 2009) are a few instances of environmental performance metrics used in project management.

The third component in evaluating sustainability is the social component. Social sustainability is given less attention than economic and environmental sustainability due to the complex explanations and challenging quantifications (Ma et al., 2018). However, as it reflects the standard of human existence and has significant effects on society, social sustainability merits considerably greater focus and effort. Benoît et al. (2007) categorized social assessment indicators into five groups: workforce, society, local community, consumers, and value-chain stakeholders. Their classification substantially benefited in the application of research on social sustainability assessments. The UNEP "Scientific and Technical Advisory Committee" (UNEP, SETAC, 2009, 2013) has provided documentation on standards and approaches regarding Social-LCA (life-cycle analysis) as a supplement to the social sustainability assessment. These documents have been further used in several studies to conduct social assessments.

### 2.2.1 UN-SDGs: Where Do We Stand at Present?

Although the "United Nations Department of Economic and Social Affairs" (UNDESA) presented 17 SDGs, all of the goals could not be addressed with the

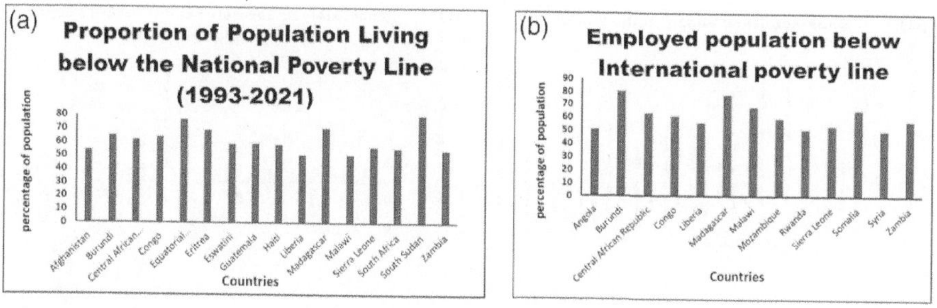


**FIGURE 2.1** (a) Countries with more than 50% population living below the national poverty line (1993–2021) and (b) countries with more than 50% employed population below international poverty line.

help of RS and MCDM. A thorough literature survey yielded the prominent sectors of SDGs which can be approached and quantifiable using RS and MCDM. A brief description of those goals is presented below.

#### 2.2.1.1 Goal 1 – No Poverty

The definition of extreme poverty is living on a purchasing power parity of less than US$2.15 per/day in 2017. However, the prevalence of this definition significantly decreased over the past few decades until the COVID-19 pandemic struck, which reversed this favourable tendency. The percentage of people living in extreme poverty dropped from 10.8% in 2015 to 8.4% in 2019. However, during the period between 2015 and 2019, the average annual decline rate was only 0.54%, which is less than half of the 1.28% rate observed between 2000 and 2014 (United Nations (UN), 2023). Figure 2.1a indicates the countries with more than 50% population living below the national poverty line and Figure 2.1b indicates countries with more than 50% employed population below the international poverty line. The data suggests that the majority of African countries are suffering from acute poverty, and employment alone can't lift them out of the poverty line. In 2020, the number of people living in extreme poverty increased by 90 million to reach a total of 724 million, reversing almost three years of progress in poverty reduction.

#### 2.2.1.2 Goal 2 – Zero Hunger

The number of people worldwide suffering chronic hunger increased by 9.2% in 2022, up from 7.9% in 2019. According to the hunger index, about 735 million people globally will be without adequate food in 2022, a shocking 122 million more than in 2019. Moreover, approximately a population of 2.4 billion or 29.6% of the global population suffered food insecurity ranging from moderate to severe level, which indicates they lacked regular access to sufficient food (United Nations (UN), 2023). While Africa has a greater proportion of its population suffering from hunger than other areas, Asia has the highest number of those suffering from hunger. More than 600 million people globally are expected to be malnourished by 2030, emphasizing the enormous difficulty of attaining the zero-hunger objective. Figure 2.2a and b indicate the most undernourished countries in the world. India

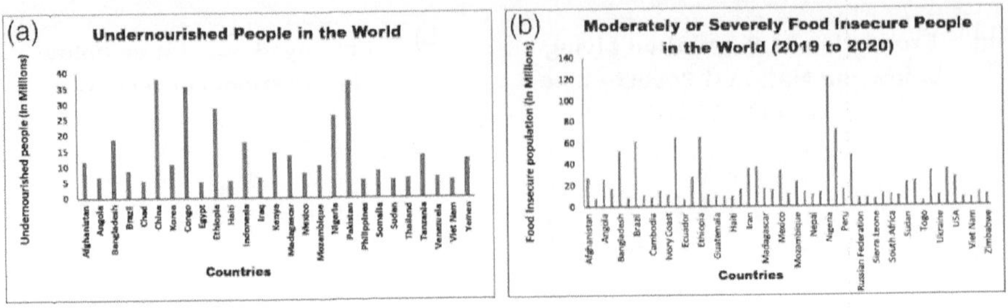


**FIGURE 2.2** (a) Countries with more than 5% undernourished population and (b) countries with more than 5% moderately or severely food insecure people.

with 224.3 million undernourished population tops the database as the country struggles to feed more than 1.4 billion people.

#### 2.2.1.3 Goal 6 – Clean Water and Sanitation

The proportion of the global population having adequate access to clean, safe drinking water rose from 49% to 57% between 2015 and 2022. Then, the proportion of people with access to safe sanitation services rose from 49% to 57%, and those with basic hygiene services rose from 67% to 75%. Figures 2.3a and b indicate the present state of drinking water and domestic wastewater treatment in third-world countries. As of 2022, a population of 2.2 billion still does not have adequate access to safe drinking water, and around 703 million lack access to basic water services (United Nations (UN), 2023). Additionally, 1.5 billion people lack access to basic sanitation, and nearly 2 billion do not possess sufficient basic household handwashing facilities, with 653 million completely devoid of any facilities.

#### 2.2.1.4 Goal 7 – Affordable and Clean Energy

From 87% in 2015 to 91% in 2021, more than 800 million more individuals will have access to power globally. In 2021, however, 675 million people who live in the least

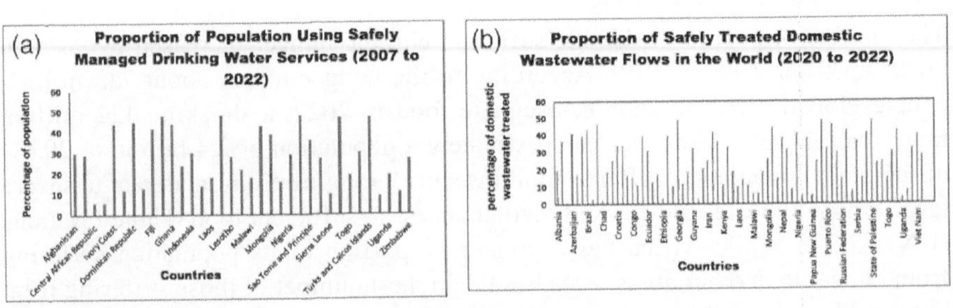


**FIGURE 2.3** (a) Countries with < 50% of the population utilizing safe drinking water services (2007–2022) and 2.3(b) countries with < 50% safely treated domestic wastewater.

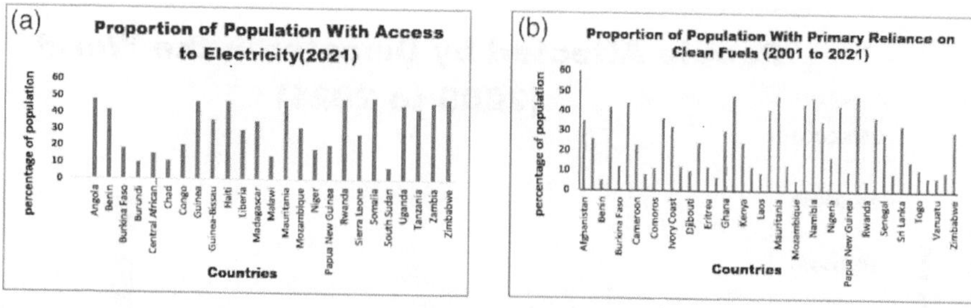


**FIGURE 2.4** (a) Countries with < 50% population with access to electricity and (b) countries with < 50% population have access to clean energy resources.

developed countries (LDCs) remained without access to electricity (Figure 2.4a). With the current trend, by 2030, 660 million people will still lack access to clean energy, especially in Africa (Figure 2.4b).

#### 2.2.1.5 Goal 11 – Sustainable Communities and Cities

In November 2022, the world's population was 8 billion; 55% of them resided in cities, and this figure is anticipated to rise to 70% by 2050. The majority of urban growth is concentrated in small- and middle-sized municipalities, which exacerbates disparities and urban poverty. Although the proportion of the urban populace living in slums has slightly decreased from 25.4% in 2014 to 24.2% in 2020, the overall slum-dweller population has still increased owing to urbanization (United Nations (UN), 2023), as shown in Figure 2.5. The number of urban residents living in slums or conditions similar to slums is estimated to reach 1.1 billion by 2020. Furthermore,

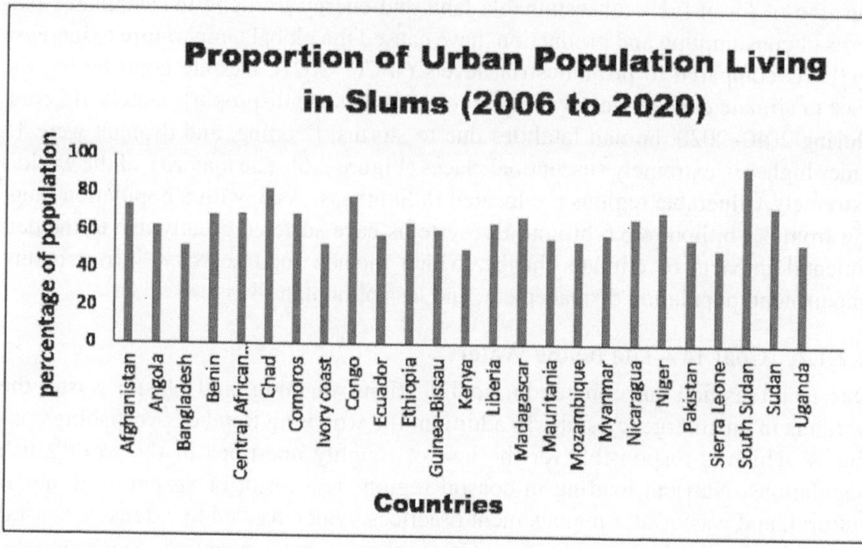


**FIGURE 2.5** Countries with more than 50% urban population living in slums.

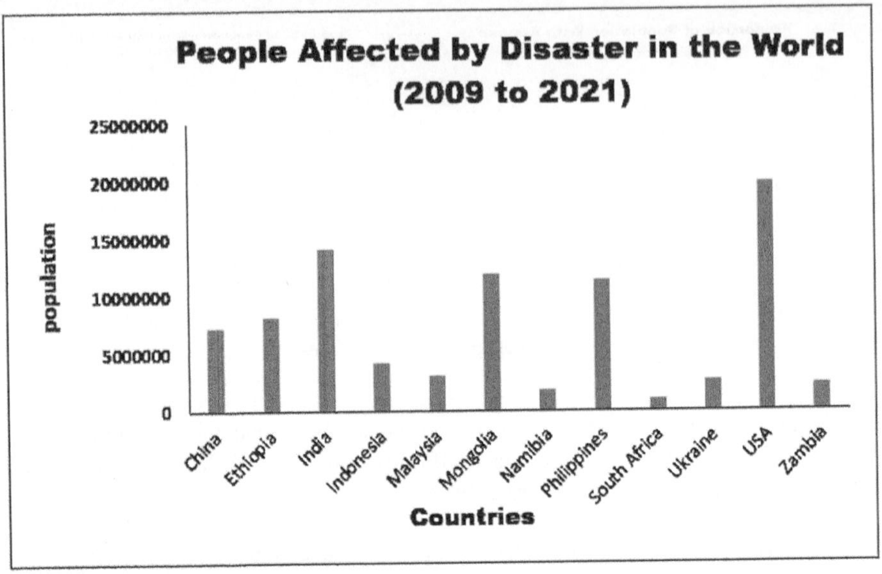


**FIGURE 2.6** Countries with more than 1 million population affected by natural disasters.

an additional population of 2 billion is projected to reside in similar settlements during the next 30 years.

### 2.2.1.6 Goal 13 – Climate Action

According to the recent synthesis report of "The Intergovernmental Panel on Climate Change" (IPCC), it is an undeniable fact that human activities, such as the burning of fossil fuels, unsustainable land and energy use, and unsustainable patterns of consumption and production, have caused the global temperature to increase by 1.1°C compared to pre-industrial levels (IPCC, 2021). Despite contributing the least to climate change, disadvantaged communities are disproportionately affected. During 2010–2020, human fatalities due to storms, flooding, and drought were 15 times higher in extremely susceptible places (Figure 2.6). The majority of the world's extremely vulnerable regions are located in Southeast Asia, with a population ranging from 3.3 billion to 3.6 billion. Ecosystems have suffered greatly due to the detrimental impacts of climate change, which include food scarcity, infrastructure impairment, population displacement, and loss of human lives.

### 2.2.1.7 Goal 14 – Life below Water

Due to intensified eutrophication, acidification, warming, and plastic waste, the ocean is in an emergency state. In addition, the worrying trend of overfishing continues, which is responsible for the loss of roughly one-third of the world's fish populations. Nutrient loading in coastal regions is a result of agricultural, aquacultural, and wastewater management practices, which has led to extensive coastal eutrophication and algal blooms. These blooms have negative consequences, including oxygen depletion, marine life destruction, contaminated aquatic creatures, and damage to coral reefs, and seagrass. Although there is an international

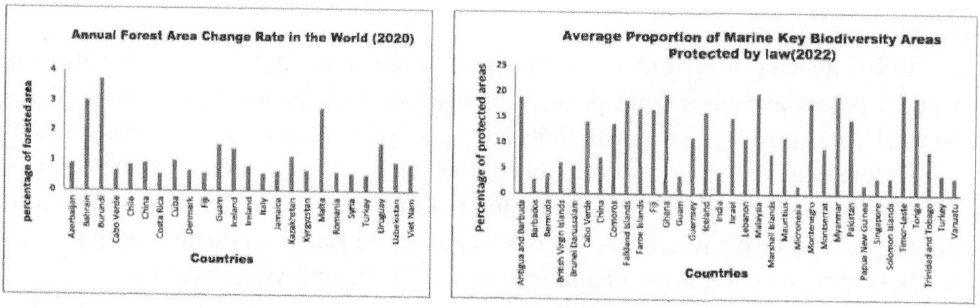


**FIGURE 2.7** (a) Countries with more than 0.5% change in forested areas and (b) countries with less than 20% marine biodiversity area protected by law.

law for protecting marine biodiversity areas, the success of implementation is very low across the globe (Figure 2.7b).

#### 2.2.1.8 Goal 15 – Life on Land

Forests are critical in inhibiting climate change and providing essential goods, services, and livelihoods. However, net forest coverage decreased by approximately 100 million hectares during the previous two decades (Figure 2.7a). Between 2000 and 2020, global forest coverage decreased by 0.7% from 4.2 billion to 4.1 billion hectares. Agricultural expansion is directly responsible for about 90% of deforestation worldwide (with 49.5% and 38.5% due to agriculture and livestock grazing). Between 2000 and 2018, deforestation worldwide was caused by the harvesting of palm oil, accounting for approximately 7% of the overall deforestation.

## 2.3 CONTRIBUTION OF MCDM TO BUILDING A RESILIENT SOCIETY

MCDM could help develop solutions to local, regional, and global spatial issues. Coupled with RS, geographic information system (GIS), and newly developed Artificial Intelligence (AI) techniques, the applicability of MCDM is tremendous. This section provides comprehensive information about the various applications of MCDM in different sectors of society.

### 2.3.1 MCDM in Poverty Alleviation

Different countries have implemented multidimensional indicators of poverty to assess different aspects of poverty, including access to essential services, employment, education, health, and the interdependence of various family objectives. Despite implementing similar policies, there has been a significant lack of progress in reducing national multidimensional poverty. MCDS could turn out to be a significant apparatus in addressing the issue of poverty and identification of robust solutions. For instance, photovoltaic poverty alleviation is one of the ways to help the marginal population become economically stable. Tao et al. (2022) applied this method in Yunnan, China, by addressing four major aspects: economy, technology, society, and environment. MCDM is also useful for determining poverty type (Dou et al., 2022) and poverty indicator for a specified region (Rubio et al., 2018). To solve the problem of assigning unique values to specific indicators, Ming et al. (2020) employed Pythagorean fuzzy sets (PFSs) instead of typical algorithms that only use a single indication valuation to shield potential values from positive and negative elements. The efficacy of RS and MCDM as poverty alleviation tools has been well-documented by various scientists. Unfortunately, the number of publications on this subject remains quite limited, particularly for third-world and developing nations where these tools are of utmost importance. Noteworthy studies on the matter have been conducted by Ke et al. (2022), Budiman et al. (2018), Olajuyigbe et al. (2013), and Ariyani (2016).

### 2.3.2 MCDM in Achieving Zero Hunger

In particular, the goal of "Zero Hunger" symbolizes a long-overdue recognition that agricultural industrialization endangers critical ecosystem processes that support food supply (Rockström et al., 2009). The consequences include biodiversity loss, increased threat of pests, soil erosion, loss of organic matter in the soil, emissions of Greenhouse Gases (GHGs), eutrophication, and pollution of water bodies (Diaz and Rosenberg, 2008; Foley et al., 2011; Matson et al., 1997). Tiwari et al. (1999) created an MCDM framework that effectively incorporates criteria of sustainability and the perspectives of local communities to facilitate precise decision-making in environmental and economic domains. Apart from site suitability analysis (Saha et al., 2021), MCDM methods could be used in agriculture for supply chain risk management (Yazdani et al., 2021), water distribution problem (Garai and Garg, 2022; Tork et al., 2021), management of irrigated agriculture (Chen et al., 2010), cost-benefit analysis (Lizot et al., 2021), etc. The advancement of RS technologies in the last three decades has made a profound impact on agriculture (Khanal et al., 2020; Wójtowicz et al., 2016). MCDM is applied with RS for land suitability analysis in hilly areas (Zolekar and Bhagat, 2015), groundwater potential zone assessment (Jhariya et al., 2016), drought vulnerability assessments (Sivakumar et al., 2021), etc.

### 2.3.3 MCDM in Providing Clean Water and Sanitation

Sewage treatment plants (STPs) around the world that employ obsolete methods are currently confronting new challenges due to rising concerns regarding sustainability, resource intensity, ecological footprint, secondary emissions, and life-cycle repercussions (Wang et al., 2012). Decision-makers have to evaluate sewage treatment systems based on reliability, affordability, sustainability, and functionality. Recent advances in MCDM approaches have aided in this robust evaluation by integrating the "Preference Ranking Organization Approach for Enrichment of Evaluations" (PROMETHEE) approach under the MCDM approach (Munasinghe-Arachchige et al., 2020). The MCDM approach has been extensively utilized in various domains such as urban wastewater management (Attri et al., 2022; Ćetković et al., 2023; Liu et al., 2020); assessing the long-term reliability of sewerage piping systems (de la Fuente et al., 2016); evaluating alternative water supplies (Savun-Hekimoğlu et al., 2021); developing sustainable sanitation (Lohman et al., 2023), and other similar applications. Additionally, MCDM is used nowadays for resource recovery from urban wastewater (Sucu et al., 2021), developing water quality indices (Akhtar et al., 2021), and emergency water provision after a natural disaster (Malekmohammadi et al., 2013).

### 2.3.4 MCDM in the Energy Sector

The energy system is critical to a country's economic and social growth as well as its inhabitants' living standards (Soytas and Sari, 2006; Yuan et al., 2008). The rising energy demand for fossil fuels has far-reaching global implications. Discharging toxic chemical contaminants, GHGs such as $CO_2$, and other airborne contaminants is a significant environmental concern (Bilgen et al., 2008; Omer, 2008). The use of MCDM in sustainable energy decision-making simply offers a way to get around the problem, and it has caught the interest of decision-makers. It is an approach based on operational assessments and decision-making for addressing complex issues comprising elevated levels of uncertainty, competing goals, disparate types of information, multiple stakes, and viewpoints. It also considers biophysical and social-economic structures' multifaceted and dynamic nature (Wang et al., 2009). In the modern day, MCDM is used for analysing energy supply in rural livelihoods (Cherni et al., 2007), sustainable energy planning and development (Pohekar and Ramachandran, 2004; Siksnelyte et al., 2018), site selection for renewable energy development (Shao et al., 2020), and managing problems (Mardani et al., 2017). MCDM has been proven to help identify optimal zones for wind energy harvesting by employing RS (Jangid et al., 2016) and mounting solar panels (Gašparović and Gašparović, 2019; Georgiou and Skarlatos, 2016).

### 2.3.5 MCDM Applications for Sustainable Cities and Communities

Urban sustainability may be seen as the planning of present and future urban projects in a way that is resource-efficient and environmentally friendly (Ali-Toudert et al., 2020). Thus, the goal is to advance economic prosperity and human well-being to mitigate and adapt to the adverse consequences of climate change. Ali-Toudert et al. (2020) devised a framework based on multi-criteria urban sustainability termed CAMSUD (Comprehensive Assessment Method for Sustainable Urban Development), which addresses three pillars of sustainability: society, economy, and ecology. The implementation of a "Clean Development Mechanism" (CDM) approach is essential for accelerating socio-economic growth and development. Numerous organizations, including the IPCC, UNFCCC ("United Nations Framework Convention on Climate Change"), and UNEP ("United Nations Environment Programme"), are aiming to minimize GHG emissions by implementing low-carbon and resource-intensive policies to attain sustainability. Ali et al. (2019) applied the MCDM and Bilan Carbone model to evaluate cost-benefit analysis from direct carbon emission reduction. RS and GIS can be used to address a variety of urban issues, such as solid waste management (Feyzi et al., 2019; Phonphoton and Pharino, 2019), landslide assessment (Akgun, 2012; Kavzoglu et al., 2014), flood vulnerability mapping (de Brito and Evers, 2016; Khosravi et al., 2019), healthcare assessment (Moreno-Calderón et al., 2020), mineral extraction and processing (Sitorus et al., 2019), construction, and civil engineering.

### 2.3.6 MCDM in Climate Change

Risk management for environmental initiatives requires addressing scientific discoveries with diversified inputs from multiple stakeholders with varying goals and priorities. In these circumstances, methodical decision-analytic approaches are the most successful means to address complicated behavioural and technical difficulties (McDaniels, 1999). Risk managers make decisions based on four major sources of information: (1) risk analysis results, (2) stakeholder preferences, (3) cost analysis, and (4) modelling and monitoring investigations.

So far, MCDM has been used for climate-related vulnerability analysis (Kim and Chung, 2013), evaluating heat stress (El-Zein and Tonmoy, 2015), water resource management (Miller and Belton, 2014), developing new policies (Greening and Bernow, 2004), etc. However, the ultimate use of MCDM could be observed in policy recommendations and formulations.

### 2.3.7 MCDM in the Ocean and Environmental Management

Environmental management approaches have changed a lot in the past three decades. Significant pollution issues were traditionally controlled by legislation, standards, and the assignment of control mechanisms in the 1970s and 1980s (Khalili and Duecker, 2013). More proactive environmental management plans have been developed since the early 1990s due to the inclusion of socio-economic elements and the need for many stakeholders' involvement in long-term environmental conservation strategies. These projects aimed to guarantee the sustainability of the economy and environment and the development of the market-based, voluntary initiatives presently gaining traction.

In the last two decades, we have witnessed several applications of MCDM in oceanographic sectors, like potential fishing zone identification (Kavadas et al., 2015), integrated coastal zone management (Garmendia et al., 2010), creating marine protected areas (Habtemariam and Fang, 2016), identification of seaweed farming potential zones (Teniwut et al., 2019), and site selection for coral reef regeneration (Mousavi et al., 2015). Apart from that, MCDM is being used to evaluate ecological potential (Ghafari et al., 2020), ecological risk assessment (Chunye and Delu, 2017; Malekmohammadi and Blouchi, 2014), ecosystem service valuation (Saarikoski et al., 2016), natural resource management (Herath and Prato, 2017; Mendoza and Martins, 2006), etc.

## 2.4 DISCUSSION

As mentioned above, MCDM constitutes one of the most dependable decision-making methods since it considers several qualitative and quantitative aspects that must be changed to reach the maximum beneficial outcome. Over the years, several researchers have created or enhanced several distinct MCDM techniques and types. These approaches differ primarily in how sophisticated their algorithms are, how they weight criteria, and how they represent preferences evaluation criteria. There are several approaches to MCDM available in the published literature. These approaches include the "Analytic Hierarchy Process" (AHP), "Analytic Network Process" (ANP), "Sum Weighted Approach" (SW), "Multi-Attribute Value Function Theory" (MAVT), "Multi-Attribute Utility Function Theory" (MAUT), "Elimination and Choice Expressing Reality" (ELECTRE), and "Technique for Order of Preference by Similarity to Ideal Solution" (TOPSIS). Compromised solutions must be sought since, in the majority of multi-criteria issues, there isn't an ideal solution that can fulfil all the criteria at once (Awasthi et al., 2011).

The aforementioned MCDA approaches and technologies could assist in making environmental choices within traditional management frameworks, but their full potential becomes apparent when combined with adaptable management. In conventional approaches, the most suitable management strategy is selected, and once implemented, its efficacy may not be routinely evaluated. However, the approach is bound to be evaluated at a certain point, and if it has been unsuccessful, an alternative approach will likely be implemented. Under this paradigm, any modification to the management method or admission of ambiguity regarding the managed system is liable to be considered a failure, and the goals or objectives that are set themselves need to be reassessed on regular intervals and bases (Linkov et al., 2006).

Adaptive management recognizes natural systems, which are inherently uncertain, and aims to minimize this uncertainty by gathering data on the governed systems. There are mainly two types of adaptive management: active and passive. Both forms involve setting goals, modelling the system, and selecting and applying management strategy. Incorporating local indigenous knowledge into adaptive management for disaster risk reduction is crucial, and MCDM provides an effective platform to achieve this. Disaster preparedness, mitigation, and adaptation can be seamlessly integrated into a viable strategy with the help of MCDM. Although most of the countries have implemented MCDM in disaster management, there is still a lack of awareness and policies in line with the Sendai Framework (Figure 2.8). In this context, it's important to bring up the case study by Kiker et al. (2001), which underlined that adaptive management had been employed as a technique because the present

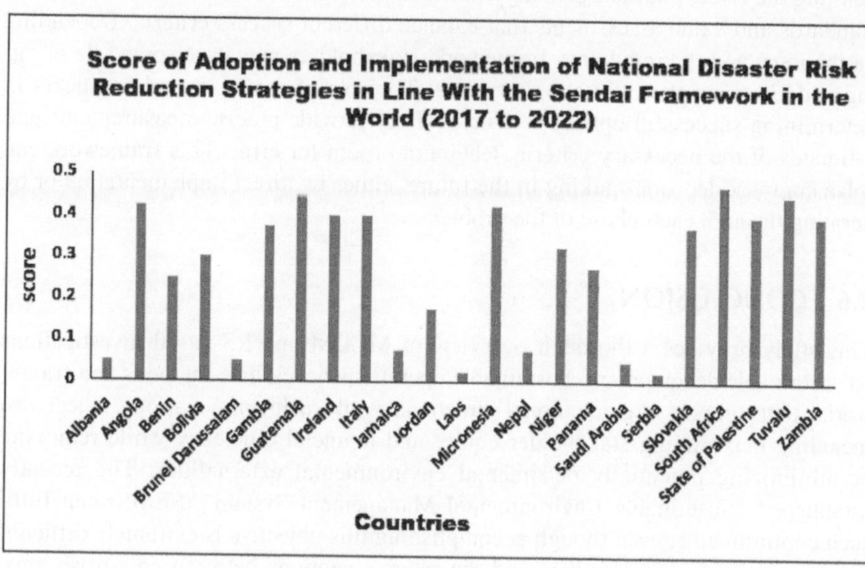


**FIGURE 2.8** Countries with score of <0.5 out of 1 in enacting national disaster risk reduction approaches in accordance with the Sendai Framework.

understanding was insufficient for ecological restoration. Later on, McDaniels and Gregory (2004) also suggested that managerial decision-making should include learning as a goal. The involvement of stakeholders is crucial to adaptive management. All stakeholders involved in the process, including the general public, businesses, academics, and governmental organizations, must feel represented and get information. Several case studies have highlighted the significance of stakeholder engagement in adaptive management (Gilmour et al., 1999; Gunderson, 1999).

## 2.5 IMPLICATIONS OF CURRENT RESEARCH

It is perceived by many scientists that MCDA in conjunction with adaptive management will offer a strong foundation for a variety of environmental management issues in the future. It will enable the making of organized, precise decisions as well as the modification of such decisions in response to their outcomes. In the future, there will be a need for a comprehensive framework to guide environmental decision-making. The first crucial element of decision-making is to have the right mix of people. Decision-makers, scientists, engineers, and stakeholders are the three key groups or divisions of people whose activity and participation levels are essential. Although the membership and responsibilities of these three categories may vary or overlap, their functions are essential to maximizing the importance of human input within the decision-making system. Each group has its unique perspective on the world, approach to problem-solving, and sense of social responsibility. The majority of the time spent by policy- and decision-makers is in identifying the context of the problem and the decision's overall restrictions. They could also be in charge of making the ultimate choice and carrying out the ensuing policy. Stakeholders can assist in defining the issue, but their primary contributions come in the form of performance standards and value assessments that balance different success criteria. Depending on the issue and the regulatory framework, stakeholders may have some role in rating and choosing the ultimate alternative. The role of scientists and engineers in determining successful options is crucial. They provide precise measurements and estimates of the necessary criteria, leaving no room for error. This framework can solve complex decision-making in the future, either by direct implementation or by iterating through each phase of the problem.

## 2.6 CONCLUSION

This study provides a thorough overview of MCDM and RS-based investigations on sustainable development. Sustainable operations necessitate methods and frameworks that support organizational functions while achieving various objectives, including maximizing stakeholder equity and financial efficiency while removing or minimizing potentially detrimental environmental externalities. The recently established "Sustainable Environmental Management System" (SEMS) can fulfil such commitments, even though accomplishing this objective is extremely difficult. Supporting long-term activities and the interconnections between enterprise, production, and planning is used to develop SEMS frameworks. Additionally, it meets the growing need for decision-making guidelines specific to the long-term reduction of environmental contaminants. In order to leverage ELECTRE III for effective decision-making, it is necessary to establish key performance benchmarks and metrics as input variables. To facilitate this, we have devised a comprehensive framework that fosters the cultivation of financial, ecological, and social standards that align with the organization's objectives and meet the expectations of both internal and external stakeholders.

## REFERENCES

Achillas, Charisios, Nicolas Moussiopoulos, Avraam Karagiannidis, Georgias Banias, and George Perkoulidis. "The use of multi-criteria decision analysis to tackle waste management problems: a literature review. " *Waste Management & Research* 31, no. 2 (2013): 115–129.

Afshari, Ali Reza, Mahdi Vatanparast, and Dragan Ćoćkalo. "Application of multi-criteria decision making to urban planning: a review." *Journal of Engineering Management and Competitiveness (JEMC)* 6, no. 1 (2016): 46–53.

Akgun, Aykut. "A comparison of landslide susceptibility maps produced by logistic regression, multi-criteria decision, and likelihood ratio methods: a case study at İzmir, Turkey." *Landslides* 9, no. 1 (2012): 93–106.

Akhtar, Naseem, Muhammad Izzuddin Syakir Ishak, Mardiana Idayu Ahmad, Khalid Umar, Mohamad Shaiful Md Yusuff, Mohd Talha Anees, Abdul Qadir, and Yazan Khalaf Ali Almanasir. "Modification of the water quality index (WQI) process for simple calculation using the multi-criteria decision-making (MCDM) method: a review." *Water* 13, no. 7 (2021): 905.

Ali, Ghaffar, Sawaid Abbas, Yanchun Pan, Zhimin Chen, Jafar Hussain, Muhammad Sajjad, and Aqdas Ashraf. "Urban environment dynamics and low carbon society: multi-criteria decision analysis modeling for policy makers." *Sustainable Cities and Society* 51 (2019): 101763.

Ali-Toudert, Fazia, Limei Ji, Linda Fährmann, and Sina Czempik. "Comprehensive assessment method for sustainable urban development (CAMSUD)-a new multi-criteria system for planning, evaluation and decision-making." *Progress in Planning* 140 (2020): 100430.

Ariyani, Nafiah. "Zakat as a sustainable and effective strategy for poverty Alleviation: from the perspective of a multi-dimensional analysis." *International Journal of Zakat* 1, no. 1 (2016): 88–106.

Aruldoss, Martin, T. Miranda Lakshmi, and Venkatasamy Prasanna Venkatesan. "A survey on multi criteria decision making methods and its applications." *American Journal of Information Systems* 1, no. 1 (2013): 31–43.

Attri, Shubham Dutt, Shweta Singh, Atul Dhar, and Satvasheel Powar. "Multi-attribute sustainability assessment of wastewater treatment technologies using combined fuzzy multi-criteria decision-making techniques." *Journal of Cleaner Production* 357 (2022): 131849.

Avtar, Ram, Akinola Adesuji Komolafe, Asma Kouser, Deepak Singh, Ali P. Yunus, Jie Dou, Pankaj Kumar et al. "Assessing sustainable development prospects through remote sensing: a review." *Remote Sensing Applications: Society and Environment* 20 (2020): 100402.

Awasthi, Anjali, Satyaveer S. Chauhan, and Hichem Omrani. "Application of fuzzy TOPSIS in evaluating sustainable transportation systems." *Expert Systems with Applications* 38, no. 10 (2011): 12270–12280.

Benoît, Catherine, Julie Parent, Isabelle Kuenzi, and Jean-Pierre Revéret. "Developing a methodology for social life cycle assessment: the North American tomato's CSR case." In *3rd International Conference on Life Cycle Management, August 2–29, Zürich, Switzerland.* 2007.

Bilgen, Selçuk, Sedat Keleş, Abdullah Kaygusuz, Ahmet Sarı, and Kamil Kaygusuz. "Global warming and renewable energy sources for sustainable development: a case study in Turkey." *Renewable and Sustainable Energy Reviews* 12, no. 2 (2008): 372–396.

Bovea, María Dolores, and Victoria Pérez-Belis. "A taxonomy of ecodesign tools for integrating environmental requirements into the product design process." *Journal of Cleaner Production* 20, no. 1 (2012): 61–71.

Boyd, Doreen S., and Francis Mark Danson. "Satellite remote sensing of forest resources: three decades of research development." *Progress in Physical Geography* 29, no. 1 (2005): 1–26.

Brones, Fabien, and Marly Monteiro de Carvalho. "From 50 to 1: integrating literature toward a systemic ecodesign model." *Journal of Cleaner Production* 96 (2015): 44–57.

Budiman, Edy, P. M. Labulan, and Ummul Hairah. "A model for poverty alleviation strategies: decision making management and indicators issue." *Advanced Science Letters* 24, no. 11 (2018): 8729–8735.

Byggeth, Sophie, and Elisabeth Hochschorner. "Handling trade-offs in ecodesign tools for sustainable product development and procurement." *Journal of Cleaner Production* 14, no. 15–16 (2006): 1420–1430.

Ćetković, Jasmina, Miloš Knežević, Radoje Vujadinović, Esad Tombarević, and Marija Grujić. "Selection of wastewater treatment technology: AHP method in multi-criteria decision making." *Water* 15, no. 9 (2023): 1645.

Chawla, Ila, Lanka Karthikeyan, and Ashok K. Mishra. "A review of remote sensing applications for water security: quantity, quality, and extremes." *Journal of Hydrology* 585 (2020): 124826.

Chen, Yun, Shahbaz Khan, and Zahra Paydar. "To retire or expand? A fuzzy GIS-based spatial multi-criteria evaluation framework for irrigated agriculture." *Irrigation and Drainage: The Journal of the International Commission on Irrigation and Drainage* 59, no. 2 (2010): 174–188.

Cherni, Judith A., Isaac Dyner, Felipe Henao, Patricia Jaramillo, Ricardo Smith, and Raúl Olalde Font. "Energy supply for sustainable rural livelihoods. A multi-criteria decision-support system." *Energy Policy* 35, no. 3 (2007): 1493–1504.

Chunye, Wang, and Pan Delu. "Zoning of Hangzhou Bay ecological red line using GIS-based multi-criteria decision analysis." *Ocean & Coastal Management* 139 (2017): 42–50.

de Brito, Mariana Madruga, and Mariele Evers. "Multi-criteria decision-making for flood risk management: a survey of the current state of the art." *Natural Hazards and Earth System Sciences* 16, no. 4 (2016): 1019–1033.

de la Fuente, Albert, Oriol Pons, Alejandro Josa, and Antonio Aguado. "Multi-criteria decision making in the sustainability assessment of sewerage pipe systems." *Journal of Cleaner Production* 112 (2016): 4762–4770.

Diaz, Robert J., and Rutger Rosenberg. "Spreading dead zones and consequences for marine ecosystems." *Science* 321, no. 5891 (2008): 926–929.

Dobrovolskienė, Nomeda, and Rima Tamošiūnienė. "Sustainability-oriented financial resource allocation in a project portfolio through multi-criteria decision-making." *Sustainability* 8, no. 5 (2016): 485.

Dou, Haojian, Llibang Ma, Shichun Liu, and Fang Fang. "Identification of rural regional poverty type based on spatial multi-criteria decision-making—taking Gansu Province, an underdeveloped area in China, as an example." *Environment, Development and Sustainability* 24, no. 3 (2022): 3439–3460.

Dube, Timothy, Munyaradzi D. Shekede, and Christian Massari. "Remote sensing for water resources and environmental management." *Remote Sensing* 15, no. 1 (2022): 18.

Eguchi, Ronald T., Charles K. Huyck, Shubharoop Ghosh, and Beverley J. Adams. "The application of remote sensing technologies for disaster management." In *The 14th World Conference on Earthquake Engineering*, vol. 17. 2008.

Elliott, Jennifer. *An introduction to sustainable development.* Routledge, 2012.

El-Zein, Abbas, and Fahim N. Tonmoy. "Assessment of vulnerability to climate change using a multi-criteria outranking approach with application to heat stress in Sydney." *Ecological Indicators* 48 (2015): 207–217.

Feyzi, Sadaf, Mehrdad Khanmohammadi, Niloofar Abedinzadeh, and Mehdi Aalipour. "Multi-criteria decision analysis FANP based on GIS for siting municipal solid waste incineration power plant in the north of Iran." *Sustainable Cities and Society* 47 (2019): 101513.

Foley, Jonathan A., Navin Ramankutty, Kate A. Brauman, Emily S. Cassidy, James S. Gerber, Matt Johnston, Nathaniel D. Mueller et al. "Solutions for a cultivated planet." *Nature* 478, no. 7369 (2011): 337–342.

Garai, Totan, and Harish Garg. "Multi-criteria decision making of water resource management problem (in agriculture field, Purulia district) based on possibility measures under generalized single valued non-linear bipolar neutrosophic environment." *Expert Systems with Applications* 205 (2022): 117715.

Garmendia, Eneko, Gonzalo Gamboa, Javier Franco, Joxe Mikel Garmendia, Pedro Liria, and Marta Olazabal. "Social multi-criteria evaluation as a decision support tool for integrated coastal zone management." *Ocean & Coastal Management* 53, no. 7 (2010): 385–403.

Gašparović, Iva, and Mateo Gašparović. "Determining optimal solar power plant locations based on remote sensing and GIS methods: a case study from Croatia." *Remote Sensing* 11, no. 12 (2019): 1481.

Georgiou, Andreas, and Dimitrios Skarlatos. "Optimal site selection for sitting a solar park using multi-criteria decision analysis and geographical information systems." *Geoscientific Instrumentation, Methods and Data Systems* 5, no. 2 (2016): 321–332.

Ghafari, Sara, Behzad Kaviani, Shahram Sedaghathoor, and Mohammad Sadegh Allahyari. "Ecological potentials of trees, shrubs and hedge species for urban green spaces by multi criteria decision making." *Urban Forestry & Urban Greening* 55 (2020): 126824.

Gilmour, Alistair, Greg Walkerden, and James Scandol. "Adaptive management of the water cycle on the urban fringe: three Australian case studies." *Conservation Ecology* 3, no. 1 (1999).

Greening, Lorna A., and Steve Bernow. "Design of coordinated energy and environmental policies: use of multi-criteria decision-making." *Energy Policy* 32, no. 6 (2004): 721–735.

Gunderson, Lance. "Resilience, flexibility and adaptive management––antidotes for spurious certitude?" *Conservation Ecology* 3, no. 1 (1999).

Habtemariam, Bereket Tesfamariam, and Qinhua Fang. "Zoning for a multiple-use marine protected area using spatial multi-criteria analysis: The case of the Sheik Seid Marine National Park in Eritrea." *Marine Policy* 63 (2016): 135–143.

Herath, Gamini, and Tony Prato. "Role of multi-criteria decision making in natural resource management." In *Using multi-criteria decision analysis in natural resource management*, pp. 17–26. Routledge, 2017.

Jangid, Jayant, Apurba Kumar Bera, Manoj Joseph, Vishal Singh, Tarun Pratap Singh, B. K. Pradhan, and Sandipan Das. "Potential zones identification for harvesting wind energy resources in desert region of India–a multi criteria evaluation approach using remote sensing and GIS." *Renewable and Sustainable Energy Reviews* 65 (2016): 1–10.

Jato-Espino, Daniel, Elena Castillo-Lopez, Jorge Rodriguez-Hernandez, and Juan Carlos Canteras-Jordana. "A review of application of multi-criteria decision making methods in construction." *Automation in Construction* 45 (2014): 151–162.

Jensen, John R. *Remote sensing of the environment: An earth resource perspective 2/e.* Pearson Education India, 2009.

Jensen, John R., and Dave C. Cowen. "Remote sensing of urban/suburban infrastructure and socio-economic attributes." *Photogrammetric Engineering and Remote Sensing* 65 (1999): 611–622.

Jhariya, Dal Chand, Tarun Kumar, M. Gobinath, Prabhat Diwan, and Nawal Kishore. "Assessment of groundwater potential zone using remote sensing, GIS and multi criteria decision analysis techniques." *Journal of the Geological Society of India* 88 (2016): 481–492.

Joyce, Karen E., Kim C. Wright, Sergey V. Samsonov, and Vincent G. Ambrosia. "Remote sensing and the disaster management cycle." *Advances in Geoscience and Remote Sensing* 48, no. 7 (2009): 317–346.

Jugend, Daniel, and José Figueiredo. "Integrating environmental sustainability and project portfolio management: case study in an energy firm." *Gestão & Produção* 24 (2017): 526–537.

Kamari, M. L., Hassan Isvand, and M. Alhuyi Nazari. "Applications of multi-criteria decision-making (MCDM) methods in renewable energy development: A review." *Renewable Energy Research and Applications* 1, no. 1 (2020): 47–54.

Kavadas, Stefanos, Irida Maina, Dimitrios Damalas, Ioannis Dokos, Maria Pantazi, and Vassiliki Vassilopoulou. "Multi-criteria decision analysis as a tool to extract fishing footprints: application to small scale fisheries and implications for management in the context of the maritime spatial planning directive." *Mediterranean Marine Science* 16, no. 2 (2015): 294–304.

Kavzoglu, Taskin, Emrehan Kutlug Sahin, and Ismail Colkesen. "Landslide susceptibility mapping using GIS-based multi-criteria decision analysis, support vector machines, and logistic regression." *Landslides* 11 (2014): 425–439.

Ke, Yiming, Hongyu Tang, Min Liu, and Xiao Qi. "A hybrid decision-making framework for photovoltaic poverty alleviation project site selection under intuitionistic fuzzy environment." *Energy Reports* 8 (2022): 8844–8856.

Khalili, Nasrin R., and Susanna Duecker. "Application of multi-criteria decision analysis in design of sustainable environmental management system framework." *Journal of Cleaner Production* 47 (2013): 188–198.

Khanal, Sami, Kushal K. C., John P. Fulton, Scott Shearer, and Erdal Ozkan. "Remote sensing in agriculture—accomplishments, limitations, and opportunities." *Remote Sensing* 12, no. 22 (2020): 3783.

Khosravi, Khabat, Himan Shahabi, Binh Thai Pham, Jan Adamowski, Ataollah Shirzadi, Biswajeet Pradhan, Jie Dou et al. "A comparative assessment of flood susceptibility modeling using multi-criteria decision-making analysis and machine learning methods." *Journal of Hydrology* 573 (2019): 311–323.

Kiker, Clyde F., J. Walter Milon, and Alan W. Hodges. "South Florida: The reality of change and the prospects for sustainability: adaptive learning for science-based policy: The Everglades restoration." *Ecological Economics* 37, no. 3 (2001): 403–416.

Kim, Yeonjoo, and Eun-Sung Chung. "Assessing climate change vulnerability with group multi-criteria decision making approaches." *Climatic Change* 121 (2013): 301–315.

Klemas, Victor. "Fisheries applications of remote sensing: An overview." *Fisheries Research* 148 (2013): 124–136.

Knight, Paul, and James O. Jenkins. "Adopting and applying eco-design techniques: a practitioners perspective." *Journal of Cleaner Production* 17, no. 5 (2009): 549–558.

Krishnaveni, K. S., and P. P. Anilkumar. "Managing urban sprawl using remote sensing and GIS." *The International Archives of the Photogrammetry, Remote Sensing and Spatial Information Sciences* 42 (2020): 59–66.

Labuschagne, Carin, and Alan C. Brent. "Sustainable project life cycle management: the need to integrate life cycles in the manufacturing sector." *International Journal of Project Management* 23, no. 2 (2005): 159–168.

Larson, Erik, and Clifford Gray. *Project management: The managerial process 6e*. McGraw Hill, 2014.

Lechner, Alex M., Giles M. Foody, and Doreen S. Boyd. "Applications in remote sensing to forest ecology and management." *One Earth* 2, no. 5 (2020): 405–412.

Linkov, Igor, F. Kyle Satterstrom, G. Kiker, C. Batchelor, T. Bridges, and E. Ferguson. "From comparative risk assessment to multi-criteria decision analysis and adaptive management: Recent developments and applications." *Environment International* 32, no. 8 (2006): 1072–1093..

Liu, Ben, Jie Tang, Zhaoyang Li, Yu Yan, and Jingshu Chen. "Optimal selection of sewage treatment technologies in town areas: a coupled multi-criteria decision-making model." *Environmental Management* 66 (2020): 709–721.

Lizot, Mauro, Flavio Trojan, and Paulo Afonso. "Combining total cost of ownership and multi-criteria decision analysis to improve cost management in family farming." *Agriculture* 11, no. 2 (2021): 139.

Lohman, Hannah A. C., Victoria L. Morgan, Yalin Li, Xinyi Zhang, Lewis S. Rowles, Sherri M. Cook, and Jeremy S. Guest. "DMsan: A multi-criteria decision analysis framework and package to characterize contextualized sustainability of sanitation and resource recovery technologie." *ACS Environmental Au* (2023), *3*(3): 179–192.

Ma, Junfeng, James D. Harstvedt, Raed Jaradat, and Brian Smith. "Sustainability driven multi-criteria project portfolio selection under uncertain decision-making environment." *Computers & Industrial Engineering* 140 (2020): 106236.

Ma, Junfeng, Gül E. Okudan Kremer, and Charles David Ray. "A comprehensive end-of-life strategy decision making approach to handle uncertainty in the product design stage." *Research in Engineering Design* 29 (2018): 469–487.

Macharis, Cathy, and Annalia Bernardini. "Reviewing the use of multi-criteria decision analysis for the evaluation of transport projects: time for a multi-actor approach." *Transport Policy* 37 (2015): 177–186.

Mahrad, Badr El, Alice Newton, John D. Icely, Ilias Kacimi, Samuel Abalansa, and Maria Snoussi. "Contribution of remote sensing technologies to a holistic coastal and marine environmental management framework: a review." *Remote Sensing* 12, no. 14 (2020): 2313.

Malekmohammadi, Bahram, and Leila Rahimi Blouchi. "Ecological risk assessment of wetland ecosystems using multi criteria decision making and geographic information system." *Ecological Indicators* 41 (2014): 133–144.

Malekmohammadi, Bahram, Mehrdad Nazariha, and Nikou Hesari. "Emergency response planning for providing drinking water in urban areas after natural disasters using multi criteria decision making methods." In *13th Edition of the World Wide Workshop for Young Environmental Scientists (WWW-YES-2013)-Urban waters: resource or risks?*, no. 07. HAL-ENPC, 2013.

Mardani, Abbas, Edmundas Kazimieras Zavadskas, Zainab Khalifah, Norhayati Zakuan, Ahmad Jusoh, Khalil M. Nor, and Masoumeh Khoshnoudi. "A review of multi-criteria decision-making applications to solve energy management problems: two decades from 1995 to 2015." *Renewable and Sustainable Energy Reviews* 71 (2017): 216–256.

IPCC, 2021: Summary for Policymakers. In: Climate Change 2021: The Physical Science Basis. Contribution of Working Group I to the Sixth Assessment Report of the Intergovernmental Panel on Climate Change [Masson-Delmotte, V., P. Zhai, A. Pirani, S.L. Connors, C. Péan, S. Berger, N. Caud, Y. Chen, L. Goldfarb, M.I. Gomis, M. Huang, K. Leitzell, E. Lonnoy, J.B.R. Matthews, T.K. Maycock, T. Waterfield, O. Yelekçi, R. Yu, and B. Zhou (eds.)]. Cambridge University Press, Cambridge, United Kingdom and New York, NY, USA, pp. 3–32, doi:10.1017/9781009157896.001.

Matson, Pamela A., William J. Parton, Alison G. Power, and Mike J. Swift. "Agricultural intensification and ecosystem properties." *Science* 277, no. 5325 (1997): 504–509.

McDaniels, Timothy L. "An analysis of the Tatshenshini-Alsek wilderness preservation decision." *Journal of Environmental Management* 57, no. 2 (1999): 123–141.

McDaniels, Timothy L., and Robin Gregory. "Learning as an objective within a structured risk management decision process." *Environmental Science & Technology* 38, no. 7 (2004): 1921–1926.

Mendoza, Guillermo A., and Helena Martins. "Multi-criteria decision analysis in natural resource management: a critical review of methods and new modelling paradigms." *Forest Ecology and Management* 230, no. 1–3 (2006): 1–22.

Miller, Kathleen A., and Valerie Belton. "Water resource management and climate change adaptation: a holistic and multiple criteria perspective." *Mitigation and Adaptation Strategies for Global Change* 19 (2014): 289–308.

Ming, Fu, Lifang Wang, and Jian Zhou. "The identification of poverty alleviation targets based on the multiple hybrid decision-making algorithms." *IEEE Access* 8 (2020): 169585–169593.

Moreno-Calderón, Alexander, Thai S. Tong, and Praveen Thokala. "Multi-criteria decision analysis software in healthcare priority setting: a systematic review." *Pharmacoeconomics* 38 (2020): 269–283.

Mousavi, Seyed Hassan, Afshin Danehkar, Mohammad Reza Shokri, Hadi Poorbagher, and Danial Azhdari. "Site selection for artificial reefs using a new combine multi-criteria decision-making (MCDM) tools for coral reefs in the Kish Island–Persian Gulf." *Ocean & Coastal Management* 111 (2015): 92–102.

Munasinghe-Arachchige, Srimali P., Isuru S. A. Abeysiriwardana-Arachchige, Himali M. K. Delanka-Pedige, and Nagamany Nirmalakhandan. "Sewage treatment process refinement and intensification using multi-criteria decision making approach: A case study." *Journal of Water Process Engineering* 37 (2020): 101485.

Olajuyigbe, Ayo Emmanuel, S. Osakpolor, and Suleiman Abdul-Azeez Adegboyega. "Assessment of quality of life using geographical information system approach for poverty alleviation decision-making." *International Journal of Sustainable Land Use and Urban Planning* 1, no. 1 (2013).

Omer, Abdeen Mustafa. "Energy, environment and sustainable development." *Renewable and Sustainable Energy Reviews* 12, no. 9 (2008): 2265–2300.

Phonphoton, Nuchcha, and Chanathip Pharino. "Multi-criteria decision analysis to mitigate the impact of municipal solid waste management services during floods." *Resources, Conservation and Recycling* 146 (2019): 106–113.

Pohekar, Sanjay D., and Muthu Ramachandran. "Application of multi-criteria decision making to sustainable energy planning—a review." *Renewable and Sustainable Energy Reviews* 8, no. 4 (2004): 365–381.

Remery, Marie, Christian Mascle, and Bruno Agard. "A new method for evaluating the best product end-of-life strategy during the early design phase." *Journal of Engineering Design* 23, no. 6 (2012): 419–441.

Rockström, Johan, Will Steffen, Kevin Noone, Åsa Persson, F. Stuart Chapin, Eric F. Lambin, Timothy M. Lenton et al. "A safe operating space for humanity." *Nature* 461, no. 7263 (2009): 472–475.

Rubio, Cecilia, María Clara Rubio, and Elena Abraham. "Poverty assessment in degraded rural drylands in the Monte Desert, Argentina. An evaluation using GIS and multi-criteria decision analysis." *Social Indicators Research* 137 (2018): 579–603.

Saarikoski, Heli, David N. Barton, Jyri Mustajoki, Hans Keune, Erik Gomez-Baggethun, and Johannes Langemeyer. "Multi-criteria decision analysis (MCDA) in ecosystem service valuation." In OpenNESS ecosystem services reference book/Potschin, M. [edit.]; et al., pp. 1–6. 2016.

Saha, Sunil, Debabrata Sarkar, Prolay Mondal, and Sanjit Goswami. "GIS and multi-criteria decision-making assessment of sites suitability for agriculture in an anabranching site of sooin river, India." *Modeling Earth Systems and Environment* 7 (2021): 571–588.

Savun-Hekimoğlu, Basak, Barbaros Erbay, Mustafa Hekimoğlu, and Selmin Burak. "Evaluation of water supply alternatives for Istanbul using forecasting and multi-criteria decision making methods." *Journal of Cleaner Production* 287 (2021): 125080.

Shanmugapriya, Palanisamy, Selvaraj Rathika, Thanakkan Ramesh, and Ponnusamy Janaki. "Applications of remote sensing in agriculture-a review." *International Journal of Current Microbiology and Applied Sciences* 8, no. 01 (2019): 2270–2283.

Shao, Meng, Zhixin Han, Jinwei Sun, Chengsi Xiao, Shulei Zhang, and Yuanxu Zhao. "A review of multi-criteria decision making applications for renewable energy site selection." *Renewable Energy* 157 (2020): 377–403.

Siksnelyte, Indre, Edmundas Kazimieras Zavadskas, Dalia Streimikiene, and Deepak Sharma. "An overview of multi-criteria decision-making methods in dealing with sustainable energy development issues." *Energies* 11, no. 10 (2018): 2754.

Sitorus, Fernando, Jan J. Cilliers, and Pablo R. Brito-Parada. "Multi-criteria decision making for the choice problem in mining and mineral processing: Applications and trends." *Expert Systems with Applications* 121 (2019): 393–417.

Sivakumar, Vidhya Lakshmi, Rakshith Radha Krishnappa, and Manoj Nallanathel. "Drought vulnerability assessment and mapping using multi-criteria decision making (MCDM) and application of Analytic Hierarchy process (AHP) for Namakkal District, Tamilnadu, India." *Materials Today: Proceedings* 43 (2021): 1592–1599.

Soytas, Ugur, and Ramazan Sari. "Energy consumption and income in G-7 countries." *Journal of Policy Modeling* 28, no. 7 (2006): 739–750.

Stuart, Venetia, Trevor Platt, and Shubha Sathyendranath. "The future of fisheries science in management: a remote-sensing perspective." *ICES Journal of Marine Science* 68, no. 4 (2011): 644–650.

Sucu, Seda, Maria O. van Schaik, Ramazan Esmeli, Djamila Ouelhadj, Timothy Holloway, John B. Williams, Peter Cruddas, D. Brett Martinson, Wei-Shan Chen, and Hans J. Cappon. "A conceptual framework for a multi-criteria decision support tool to select technologies for resource recovery from urban wastewater." *Journal of Environmental Management* 300 (2021): 113608.

Tabor, Joseph A., and Charles F. Hutchinson. "Using indigenous knowledge, remote sensing and GIS for sustainable development." (1994).

Tao, Yao, Yunna Wu, Man Wu, Xu Luo, Feiyang He, Rui Gao, and Lihui Zhang. "Multi-criteria decision making for comprehensive benefits assessment of photovoltaic poverty alleviation project under sustainability perspective: a case study in Yunnan, China." *Journal of Cleaner Production* 346 (2022): 131175.

Teniwut, Wellem, Marimin Marimin, and Taufik Djatna. "GIS-based multi-criteria decision making model for site selection of seaweed farming information centre: a lesson from small islands, Indonesia." *Decision Science Letters* 8, no. 2 (2019): 137–150.

Tiwari, Dirgha N., Rainer Loof, and Guna Nidhi Paudyal. "Environmental–economic decision-making in lowland irrigated agriculture using multi-criteria analysis techniques." *Agricultural Systems* 60, no. 2 (1999): 99–112.

Tork, Hamed, Saman Javadi, and S. Mehdy Hashemy Shahdany. "A new framework of a multi-criteria decision making for agriculture water distribution system." *Journal of Cleaner Production* 306 (2021): 127178.

UNEP, SETAC. "Guidelines for social life cycle assessment of products: Social and socio-economic LCA guidelines complementing environmental LCA and life cycle costing, contributing to the full assessment of goods." *United Nations Environment Programme, Paris* (2009).

UNEP, SETAC. "The methodological sheets for subcategories in social life cycle assessment (S-LCA)." *United Nations Environment Program SETAC Life Cycle Initiative United Nations Environment Programme, Paris* (2013).

United Nations (UN). "The sustainable development goals report 2023: Special edition." (2023).

Wang, Jiang-Jiang, You-Yin Jing, Chun-Fa Zhang, and Jun-Hong Zhao. "Review on multi-criteria decision analysis aid in sustainable energy decision-making." *Renewable and Sustainable Energy Reviews* 13, no. 9 (2009): 2263–2278.

Wang, Xu, Junxin Liu, Nan-Qi Ren, Han-Qing Yu, Duu-Jong Lee, and Xuesong Guo. "Assessment of multiple sustainability demands for wastewater treatment alternatives: a refined evaluation scheme and case study." *Environmental Science & Technology* 46, no. 10 (2012): 5542–5549.

WCED. World commission on environment and development. Our Common Future, 17, no. 1 (1987): 1–91.

Wójtowicz, Marek, Andrzej Wójtowicz, and Jan Piekarczyk. "Application of remote sensing methods in agriculture." *Communications in Biometry and Crop Science* 11, no. 1 (2016): 31–50.

Yang, Xiaojun X., ed. *Urban remote sensing: Monitoring, synthesis and modeling in the urban environment.* John Wiley & Sons, 2021.

Yap, Jeremy Yee Li, Chiung Chiung Ho, and Choo-Yee Ting. "A systematic review of the applications of multi-criteria decision-making methods in site selection problems." *Built Environment Project and Asset Management* 9, no. 4 (2019): 548–563.

Yazdani, Morteza, Ernesto D. R. S. Gonzalez, and Prasenjit Chatterjee. "A multi-criteria decision-making framework for agriculture supply chain risk management under a circular economy context." *Management Decision* 59, no. 8 (2021): 1801–1826.

Yuan, Jia-Hai, Jian-Gang Kang, Chang-Hong Zhao, and Zhao-Guang Hu. "Energy consumption and economic growth: evidence from China at both aggregated and disaggregated levels." *Energy Economics* 30, no. 6 (2008): 3077–3094.

Yuan, Qiangqiang, Huanfeng Shen, Tongwen Li, Zhiwei Li, Shuwen Li, Yun Jiang, Hongzhang Xu et al. "Deep learning in environmental remote sensing: achievements and challenges." *Remote Sensing of Environment* 241 (2020): 111716.

Zavadskas, Edmundas Kazimieras, Jurgita Antuchevičienė, and Oleg Kapliński. "Multi-criteria decision making in civil engineering. Part II–applications." *Engineering Structures and Technologies* 7, no. 4 (2015): 151–167.

Zolekar, Rajendra Bhausaheb, and Vijay Shivaji Bhagat. "Multi-criteria land suitability analysis for agriculture in hilly zone: Remote sensing and GIS approach." *Computers and Electronics in Agriculture* 118 (2015): 300–321.

# 3 Enhancing Sustainable Development Goals (SDG) through Geographic Information Systems and Multi-Criteria Decision-Making

## *A Case Study of Hong Kong*

*Zaw Min Htet, Subhajit Ghosh, and Indrajit Pal*

## 3.1 INTRODUCTION

Intensification of natural hazards and enhanced impacts of climate change affects many metropolitan societies, especially in the Asia Pacific region which have emphasized resilience promotion for sustainable development. Hong Kong is a highly populated developing city located as a special administrative region (SAR) on the southeastern coast of China. Hong Kong is susceptible to compound floods (Lai et al., 2023) from heavy precipitation and extreme sea levels caused by tropical cyclones (TCs). The total yearly precipitation record in Hong Kong is 2400 mm, with 4–5 extreme precipitation days with over 100 mm a day every year. Hong Kong expands up to 1110.18 km$^2$, with more than half of them being hilly areas. Much of the urbanized land is in low-lying flood-prone areas, making Hong Kong highly vulnerable to flooding and other hydrological hazards, resulting in substantial economic losses and fatalities due to TC-associated floods. As reported by HKO, the mean sea level in Victoria Harbour increased by 31 mm/decade in the last 20 years. These changes extensively exacerbate floods of all types (e.g., flash, pluvial, and coastal floods), challenging Hong Kong flood risk management profoundly.

Geographic Information Systems (GIS) has been one of the primarily preferred tools in assisting and monitoring sustainable development goals (SDGs) (Avtar et al., n.d.). Many geospatial approaches have been implemented in multiple development schemas of regional, national, and multi-national scales due to their mapping, efficient data management, regional planning, visualization, policy planning, improved management, and resource allocations. The comprehension of socio-economic clusters in a spatial approach using GIS has helped national and local stakeholders to

DOI: 10.1201/9781032683928-3

plan and implement the policies to reach SDGs. GIS play a crucial role in achieving SDG 11: sustainable cities and communities. The integration of geospatial information such as land use, land cover, statistical data, and administrative and population density can be a foundation approach, to comprehend the industrial or socio-economic facilities required to be monitored, in achieving SDG 11. It is also common that many urban planners and local stakeholders use GIS to analyze and identify how the cities have been, are, and will be in the future by setting milestones or schemas under different scenarios to help keep the cities and communities sustainable.

GIS have been employed in the climate action plans in the multiple scales for SDG 13: climate action. GIS and Remote Sensing technology are a few of the primary technical supports to monitor the current climatic systems of the region. Over the years, the collection of climate data has been helpful in not only the present climate scenarios but also the likelihood of future scenarios being impacted by climate change with different carbon emissions for planning and management in advance. Understanding the multiple dimensions of the climate landscapes, major stakeholders can come together to plan, promote, finance, and implement necessary action plans as a mitigation measure.

Due to the complexity of the global challenges, SDGs have been set and are being approached in a multi-dimensional multi-disciplinary manner. Therefore, Multiple Criteria Decision-Making (MCDM) has become very popular in recent years. Within the 2030 Agenda framework, there were 143 published scientific articles between 2016 and 2020 in which MCDM was employed to understand the SDGs due to their flexibility for stakeholders to make decisions, considering all the criteria of different levels of importance for different scales of the issues. Addressing the SDG 7: affordable and clean energy, hybrid MCDM, and fuzzy MCDM approaches were the most commonly applied, followed by Analytical Hierarchy Process (AHP) and fuzzy AHP, based on the five decision problems: source selection, location, sustainability, project performance, and technological performance. The popularity of AHP in supporting decision-making for sustainable development has increased among other MCDM approaches in recent years when categorized as follows: preference modeling, uncertainty approaches, sensitivity analysis, long-term assessment, and stakeholder involvement. Multiple Criteria Decision-Making within the 2030 Agenda Framework has been implemented the most at the national level, followed by local and regional levels. Addressing SDG 7, SDG 9, and SDG 11, goal programming (GP) was used for resource mobilization, decentralized electricity generation, and new transport modes for the United Arab Emirates (UAE). Regarding SDG 11 and SDG 13, fuzzy TOPSIS was employed for sustainable transportation systems in India. AHP was implemented for SDG 4, SDG 5, and SDG 17 in Europe for the aspects of governance, civic engagement, gender equality, and more. There were cases of single MCDM, combinations of MCDM and non-MCDM methods, and integration of MCDM methods in approaching the SDGs of different scales.

## 3.2 STUDY AREA

The study area, Hong Kong Island, is the southern part and at the geographic coordinates of 22.2988°–22.2070° north latitude and 114.1732° and 114.1631° east longitude, of the Hong Kong Synthetic Aperture Radar (SAR) in China. It is located to the south of the Kowloon Peninsula and is separated from it by Victoria Harbour, bordering with Guangdong Province of Mainland China. The study area covers an area of 78.64 km$^2$ with four major districts: Central and Western, Wan Chai, Eastern, and Southern. The topography of Hong Kong Island is defined by hills in the central and coastal areas along the shore. Hong Kong Island is populated mostly along the coastal areas, especially in the northern, western, and southern parts of the island. The majority of the settlement is in the low-lying flood-prone areas, making Hong Kong challenging for sustainable development. Hong Kong is susceptible to compound floods from heavy precipitation and extreme sea levels caused by TCs. May, June, July, August, and September have monthly rainfall levels exceeding 300 mm; about 80% of the yearly rainfall occurs in a five-month period. HKO monthly mean precipitation data from 1981 to 2010 indicates that June is the rainiest month in terms of both total and duration of rainfall. It is reported by the Hong Kong Observatory Headquarters that Hong Kong recently received record-breaking rainfall per hour on September 7, 2023, with 158.1 mm, followed by 145.5 mm, 115.1 mm, 109.9 mm, and 108.2 mm on June 7, 2008, July 16, 2006, May 8, 1992, and June 12, 1966, respectively. The intensification was in effect for more than seven hours, indicating Hong Kong is susceptible to flooding (Figure 3.1).

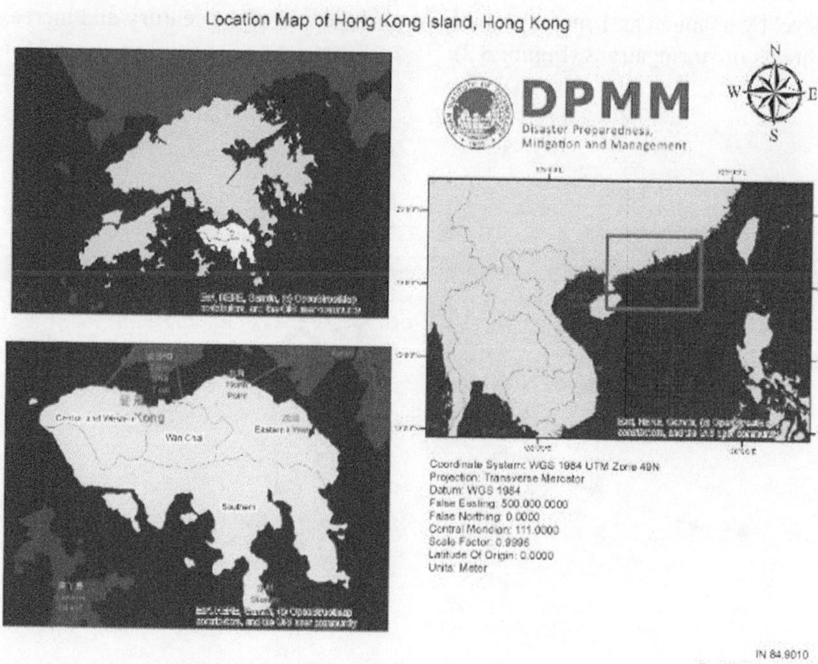


**FIGURE 3.1** Location map of Hong Kong Island, Hong Kong SAR, China.

## 3.3 CLIMATE HAZARDS IN HONG KONG

### 3.3.1 Typhoons

Hong Kong's typhoon season spans from May to November, peaking during the summer months of June, July, and August. Hong Kong has engaged with 13 natural hazards in the past 20 years (2002–2022) according to the Emergency Events Database (EM-DAT) in which 12 of them are TCs: Dujuan, Imbudo, Typhoon "Hagupit" (Nina), Tropical Storm "Kammuri" (Julian), Tropical Storm "Nuri" (Karen), Typhoon "Koppu", Typhoon Chanthu, Typhoon "Hato", Tropical Storm "Pakhar"/"Jolina", Typhoon Mangkhut (Ompong), TC "Lionrock" (Lannie), and TC "Kompasu" (Maring). The typhoons are also one of the many triggering factors for floods and landslides. The highest water level recorded by tide gauges is 1.77 m according to the HKO's database on storm surge records between 1945 and 2015. The devastation caused by Typhoon Mangkhut in 2018 highlighted the destructive impact of compound floods. The combination of heavy precipitation and extreme sea levels during the typhoon resulted in over 400 injuries and extensive economic losses.

### 3.3.2 Sea Level Rise

The major impacts of sea level rise (SLR) to Hong Kong Island would be a possible increase in frequency and scale of sea flooding from storm surges brought about by TCs (Lee et al., 2010). Hong Kong Island, where the majority of the region is low-lying areas, is prone to sea level rise as it was predicted that climate change may raise sea level by as much as 1 meter globally toward the end of the century and increase the threats of storm surges (Figure 3.2).

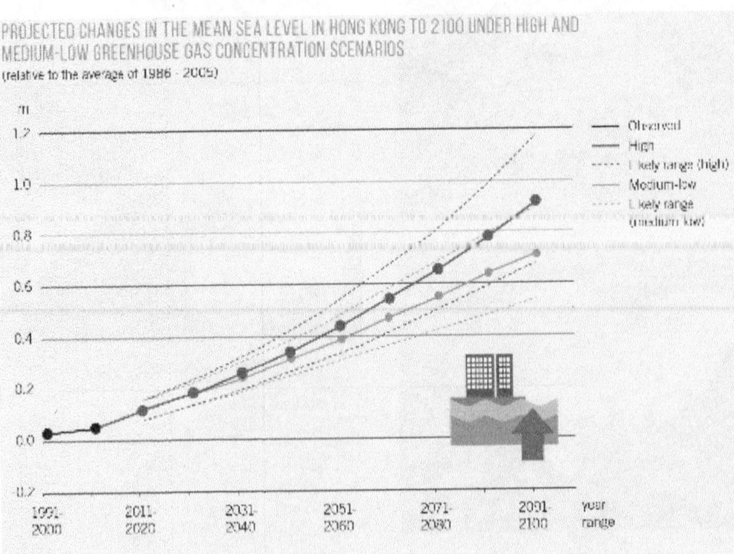


**FIGURE 3.2** Sea level rise projection of Hong Kong. (Source: Environment Bureau, 2017)

### 3.3.3 Climate Change Impacts

According to the Hong Kong Climate Change Report, climate change will lead to warmer days and warmer nights, fewer rainy days with increased average rainfall intensity, more extreme rainfall events, more extreme wet years with risk of extreme dry years, and increased potential of storm surges associated with TCs. Mean sea level in Victoria Harbour (Johnson et al., 2016) has been increasing at a rate of 30 mm in a ten-year period from 1954 to 2014. Hong Kong is projected to rise by 62 to 70–73 to 91 cm by 2100, depending on the climate change setting, relative to the 1986–2005 average sea level.

## 3.4 METHODOLOGICAL OUTLINES FOR HAZARD AND RESILIENCE STUDY

Multiple methods have been employed in studying hazards to improve regional resilience in the disaster risk management context. According to Bera et al. (2022), AHP, Shannon's Entropy (SE), and Support Vector Machine (SVM) are used to identify the flood susceptibility in coastal districts of India. Likewise, the AHP is used as a multi-criteria analysis of how parameters like elevation, drainage density, slope, and land use/cover contribute to the flooding hazards in the Sekondi-Takoradi Metropolis (STM) (Danso et al., 2024). SE is one measure of information measure using probability theory. It is employed to study how different or diverse each element is in the assessment matrix. The entropy values are then calculated to represent the degree of uncertainty in the attributes to comprehend the flood susceptibility index using the weighted sum approach.

Understanding the flood hazard context in the Eastern Hindukush region of Pakistan (Hussain et al., 2023), two machine learning models: SVM and Random Forest (RF) and AHP are used. SVM is a supervised machine learning approach, based on the principles of statistical learning theory. It is practiced to transform nonlinear (LN) data into two distinct classes, using kernel functions such as LN, Radial Basis Function (RBF), Sigmoid (SIG), and Polynomial (PL). The RBF kernel is one of the more practiced functions due to its less data-intensive approach. SVM is also being used in understanding the risk of natural hazards for SDGs.

The methodology carried out in this study is based on the comprehensive multi-hazard risk assessment in the GIS environment with the integration of multiple criteria decision analysis, AHP. This approach enables the consideration and weighting of various influencing factors to generate a robust risk assessment for building resilience. Four major elements are considered: hazard, exposure, sensitivity, and capacity.

### 3.4.1 Hazard Indicators

Hazards refer to the natural occurrence of incidents, triggered by the earth climate systems. In this study, hazard indicators are derived from the historical hazard records in Hong Kong Island in the last 20 years. TC tracks and sea level rise are both very common and crucial phenomena to derive the hazard index for risk quantification in the multi-hazard risk assessment calculation (Figure 3.3).

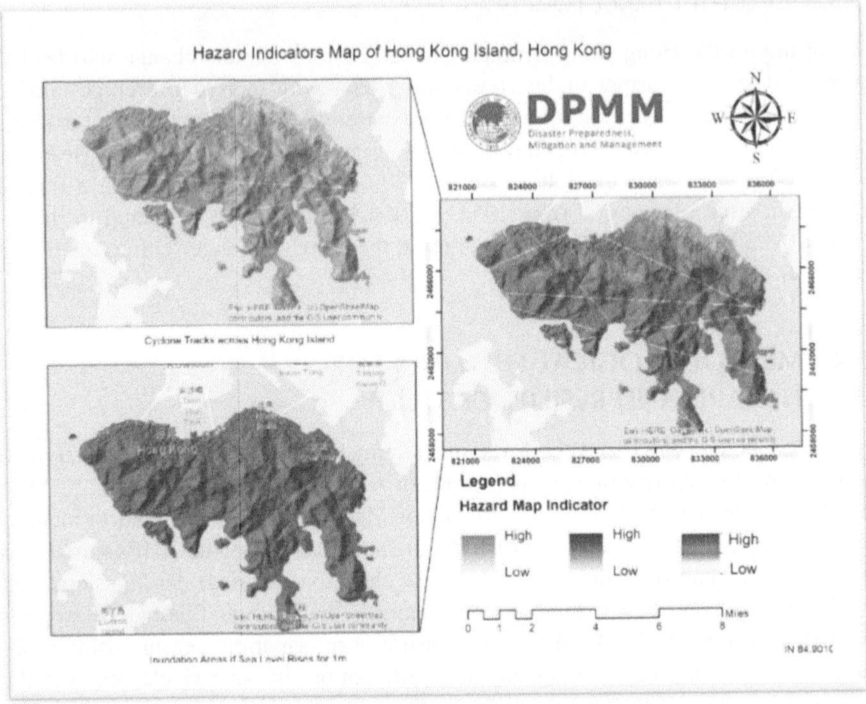


FIGURE 3.3 Hazard indicator map of Hong Kong Island.

### 3.4.2 Exposure Indicators

Exposure refers to the extent to which an area is in contact with a perturbation. The exposure component comprises the demographic density and topographic landscapes. The overall population density of Hong Kong Island is considered as the majority of the settlements are concentrated on the low-lying flood-prone areas of the island. Topographic landscape can influence the likelihood of impacts on the assets and population from disasters. The understanding of physical exposure plays a major role in integrated hazard risk assessment (Figure 3.4).

### 3.4.3 Sensitivity Indicators

Sensitivity refers to the degree of vulnerability from incidents. Population under the age of 15 and those over 65 (Sun et al., 2017) are considered highly sensitive due to their high dependency and need for assistance from others during and after the disasters. Moreover, the proximity to the shoreline is another important sensitivity indicator in this study due to the high frequency and intensification of TCs and other coastal hazards. Besides, road network density is regarded as a sensitivity indicator due to its importance for emergency response and evacuation modes of transportation amid the disaster. Land use is also considered a sensitivity indicator as different categories of land use can be affected on a different scale from natural hazards.

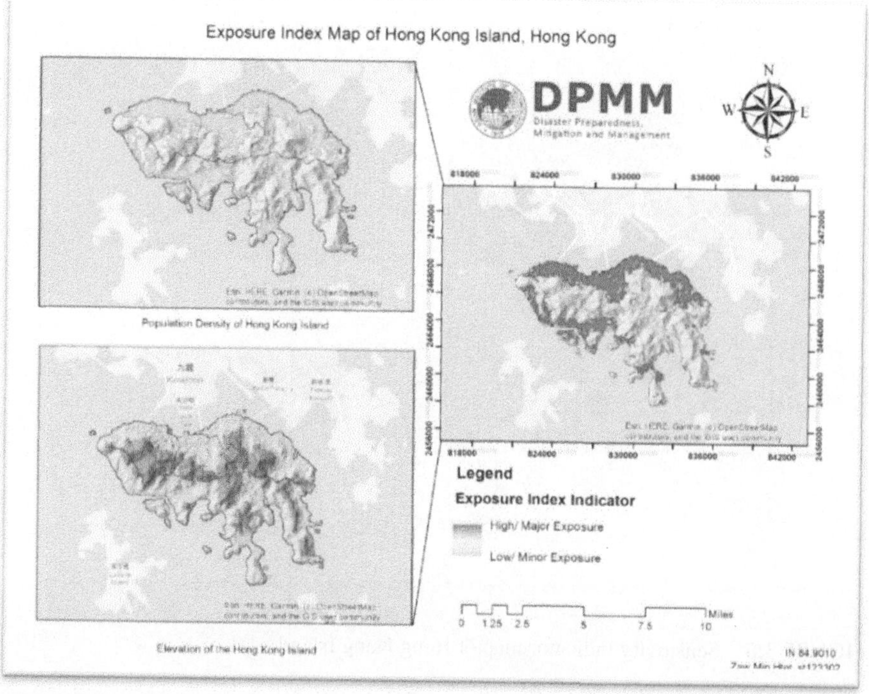


FIGURE 3.4 Exposure indicator map of Hong Kong Island.

Recognizing the sensitivity components is crucial to building resilience for sustainable developments through indicator development (Figures 3.5 and 3.6).

### 3.4.4 Capacity Indicators

Capacity indicators are crucial elements in quantifying the ability of a community, region, or system to absorb, resist, or recover from natural hazards. Median monthly income as an economic stability is considered as a primary capacity indicator. The financial ability of an individual household to respond to natural hazards plays an important role for post-disaster reconstruction actions. Moreover, secondary education as education attainment is also a critical capacity component. Educated individuals have the capacity to engage in disaster-related information to be involved in both proactive disaster preparedness activities and reactive disaster response and recovery actions. Capacity indicators, shown in Figure 3.7 are the contributing factors in minimizing the vulnerability and promoting community resilience.

### 3.4.5 Multi-Criteria Analysis: Analytical Hierarchy Process (AHP)

AHP is an effective pairwise comparison matrix in transforming attributes of the expert rankings into numerical values for quantification of the factors of importance

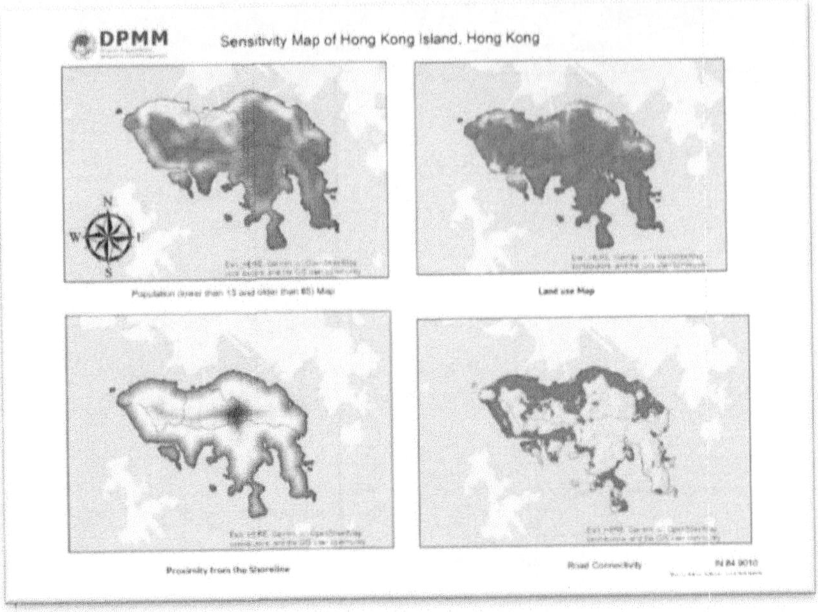


FIGURE 3.5 Sensitivity indicator map of Hong Kong Island.

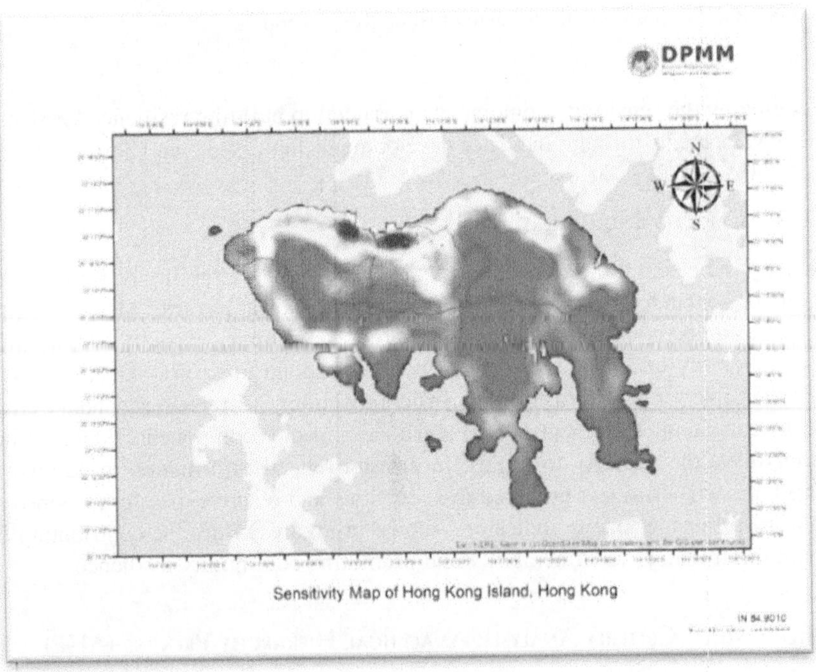


FIGURE 3.6 Overall sensitivity indicator of Hong Kong Island.

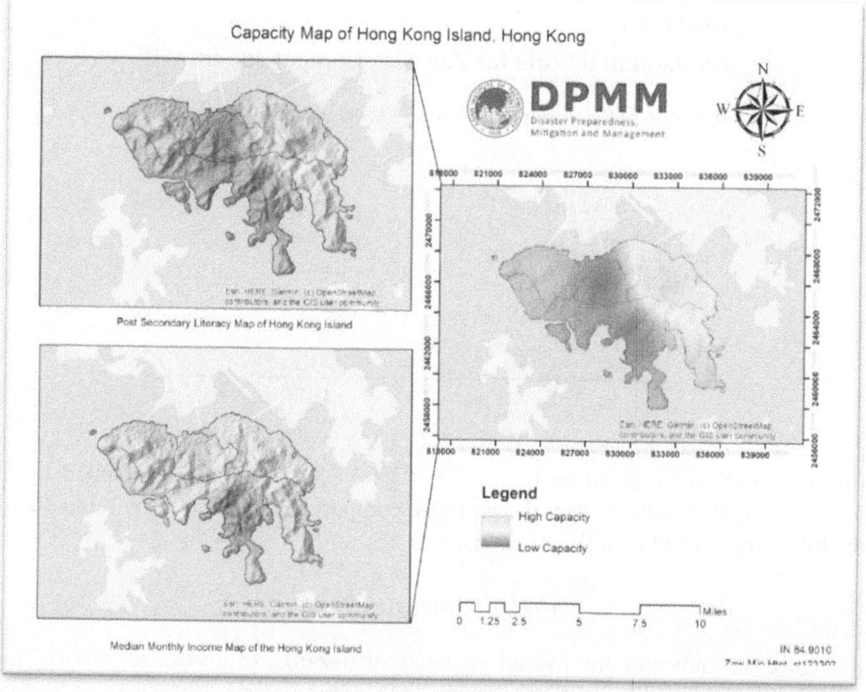


**FIGURE 3.7** Capacity indicator map of Hong Kong Island.

(Table 3.1), influencing the hazard incidents in the study area. The weights were attained by the principal eigenvector of a square reciprocal 9 × 9 matrices of pair-wise comparison of components (Bera et al., 2022). The consistency ratio (CR) was calculated to check the consistency of the determined weights using the following method.

$$\mathrm{CR} = \frac{CI}{RI}.$$

CR stands for the Consistency Ratio, derived from the calculation of CI (Consistency Index) and RI (Random Inconsistency Index) in the pairwise matrix. CR can be calculated based on the calculation below.

$$\mathrm{CI} = \frac{\gamma\text{max-n}}{n-1},$$

where $\gamma$max refers to the maximum eigenvalue of the judgment matrix and n is the total number of elements. The CR being 0.0925 confirmed the high consistency of the rating used in the study. Following the weightage calculation, a multi-hazard risk assessment map is generated, using the weighted sum approach in the Geographic

**TABLE 3.1**
**Fundamental Scale for Pairwise Comparison in AHP**

| The Scale of Relative Importance | Level of Importance |
|---|---|
| 1 | Equal |
| 3 | Moderate |
| 5 | Strong |
| 7 | Very strong |
| 9 | Extreme |
| 2, 4, 6, 8 | Intermediate |

*Source:* Danso et al. (2024).

Information System based on Figure 3.8. where each element is multiplied by its factors' weight, resulting from the pairwise comparison calculation of Table 3.2 by the following formula (Bollin et al., n.d.):

$$MHRA = (H + E + S) - C.$$

Figure 3.9 indicates the overall elements of hazards, exposure, sensitivity, and capacity of the multi-hazard assessment, highlighting a broad overview of spatial distribution of risks imposing on the Hong Kong Island, Hong Kong.

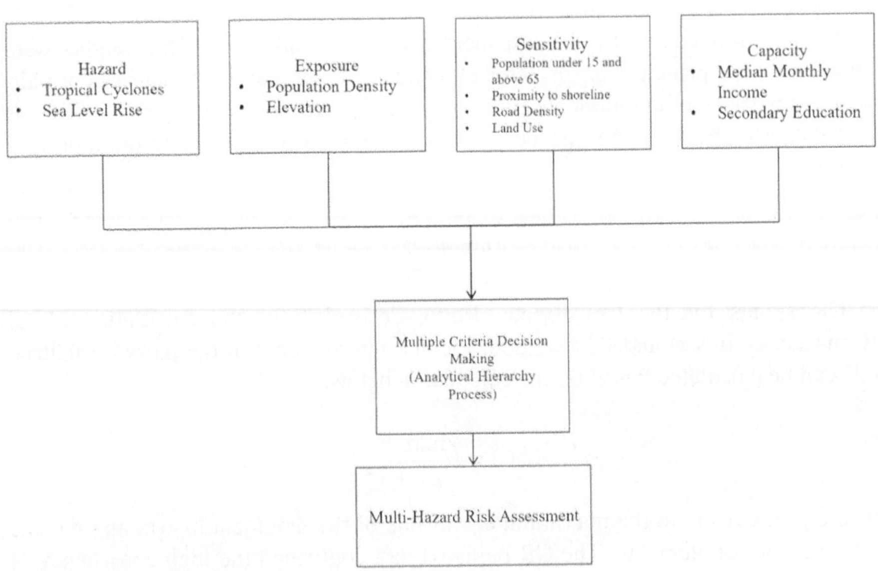


**FIGURE 3.8** Methodology.

**TABLE 3.2**
**Pairwise Comparison Consistency Ratio**

| | CT | SLR | PD | Elev | Pop | Prox | LU | RD | MMI | SE | Sum Row | Lamda Factor | Lamda Individual | Consistency Index | Consistency Ratio |
|---|---|---|---|---|---|---|---|---|---|---|---|---|---|---|---|
| **CT** | 1 | 3 | 3 | 4 | 4 | 5 | 6 | 7 | 8 | 8 | 2.941 | 34.33313 | 11.67289 | 0.137936 | 0.092574 |
| **SLR** | 1/3 | 1 | 2 | 3 | 4 | 4 | 5 | 6 | 7 | 7 | 1.947 | 23.74575 | 12.19295 | | |
| **PD** | 1/3 | 1/3 | 1 | 3 | 4 | 4 | 5 | 6 | 7 | 8 | 1.697 | 20.91023 | 12.32044 | | |
| **Elev** | 1/4 | 1/3 | 1/3 | 1 | 2 | 3 | 4 | 4 | 5 | 6 | 1.049 | 12.68446 | 12.09562 | | |
| **Pop** | 1/4 | 1/4 | 1/4 | 1/3 | 1 | 2 | 5 | 6 | 7 | 7 | 1.004 | 11.63741 | 11.59249 | | |
| **Prox** | 1/5 | 1/7 | 1/7 | 1/4 | 1/2 | 1 | 3 | 6 | 7 | 8 | 0.812 | 9.028292 | 11.11583 | | |
| **LU** | 1/6 | 1/5 | 1/5 | 1/4 | 1/5 | 1/3 | 1 | 2 | 3 | 4 | 0.415 | 4.432113 | 10.67933 | | |
| **RD** | 1/7 | 1/6 | 1/6 | 1/6 | 1/4 | 1/6 | 1/6 | 1 | 4 | 5 | 0.371 | 3.738515 | 10.07548 | | |
| **MMI** | 1/8 | 1/7 | 1/7 | 1/5 | 1/7 | 1/7 | 1/3 | 1/4 | 1 | 3 | 0.227 | 2.296014 | 10.09801 | | |
| **SE** | 1/8 | 1/7 | 1/8 | 1/6 | 1/7 | 1/8 | 1/4 | 1/5 | 1/3 | 1 | 0.160 | 1.691506 | 10.57118 | | |

*Notes:* CT = Cyclone Track, SLR = Sea Level Rise, PD = Population Density, Elev= Elevation, Pop = Population under 15 and above 65, Prox= Proximity to Shoreline, LU= Land use, RD = Road Density, MMI = Median Monthly Income, SE = Secondary Education

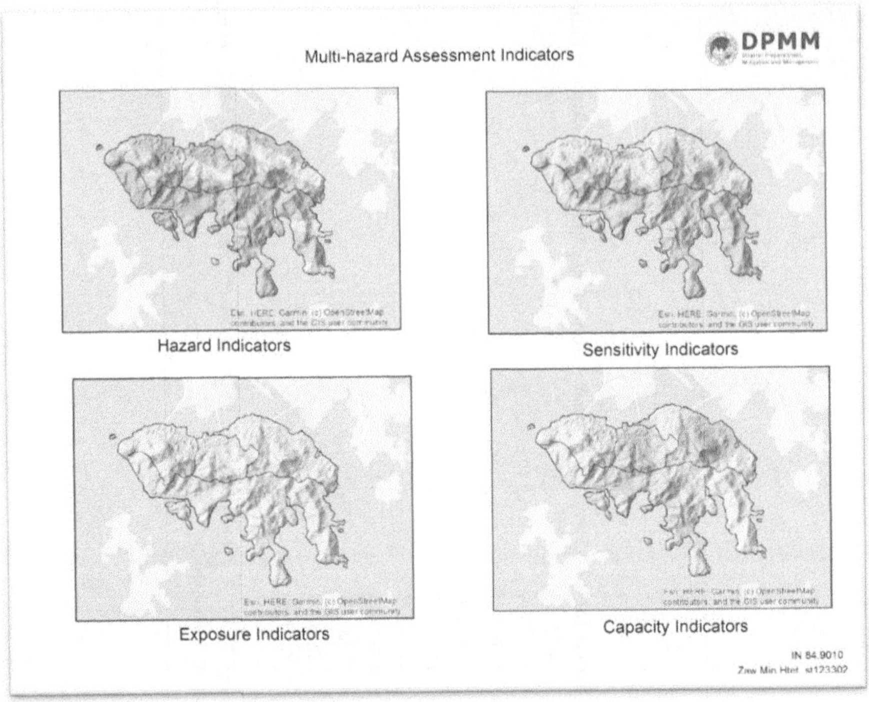


**FIGURE 3.9** All indicators for the multi-hazard risk assessment.

## 3.5 RESULT AND DISCUSSION

The multi-hazard risk assessment of Hong Kong Island is employed using the elements in Table 3.3. The hazard profile of Hong Kong indicates that the western and southern sides of the island are relatively higher to be affected by natural hazards: TCs and SLR based on the historical hazard records. The economic losses and the lives affected by TCs exceed the losses and lives affected by the SLR. The cyclone tracks recorded in the historical data imply that the Central and Western Districts and the Southern Districts have been affected by TCs multiple times in recent years.

In terms of exposure, Hong Kong Island population density is the highest in the low-lying areas especially in the Central and Western Districts, along the shoreline areas. Moreover, the road connectivity of Hong Kong Island is highly concentrated in the Central and Western District, compared to the other districts. Therefore, the majority of the socio-economic infrastructures and the most vulnerable age groups of people are exposed to natural hazards while only very few in the central mountainous topography are less exposed. The capacity indicators imply that the Eastern District is relatively higher than the other districts in terms of both financial stability and educational attainment.

The overall multi-hazard risk index indicates that socio-economic activities in the Eastern District are ranked at very high risk from multiple hazards with an estimated

**TABLE 3.3**
**Multi-Hazard Risk Assessment Indicator Matrix**

| | **Risk Assessment** | | | | |
|---|---|---|---|---|---|
| 1. | Hazard Indicators | | | | 34% |
| | **Indicator Name** | **Description** | **Data Source** | **Weights (%)** | |
| 1.1 | Cyclone tracks | Cyclone tracks going across Hong Kong Island | Secondary data | 65 | |
| 1.2 | Sea level rise | Potentially inundated areas when sea level rises by 1 m | Extracted from elevation | 35 | |
| 2. | Exposure | | | | 33% |
| | **Indicator Name** | **Description** | **Data Source** | **Weights (%)** | |
| 2.1 | Population density | Overall population density of Hong Kong Island | WorldPop (2020) | 60 | |
| 2.2 | Elevation | Elevation of the Hong Kong Island | SRTM | 40 | |
| 3. | Sensitivity | | | | 33% |
| | **Indicator Name** | **Description** | **Data Source** | **Weights (%)** | |
| 3.1 | Population<15 and population<65 | Vulnerable age group less than 15 and older than 65 | WorldPop (2020) | 20 | |
| 3.2 | Proximity | Proximity from the shoreline | Extracted from boundary | 20 | |
| 3.3 | Road density | The connectivity in between the communities of Hong Kong | Open Street Map | 30 | |
| 3.4 | Land use | Land use planning of Hong Kong | ESRI (2024) | 30 | |
| 4. | Capacity | | | | |
| | **Indicator Name** | **Description** | **Data Source** | **Weights (%)** | |
| 4.1 | Median monthly income | The financial capacity of coping with the disaster | Secondary data | 50 | |
| 4.2 | Secondary education | The intellectual capacity to cope with the disaster | Secondary data | 50 | |
| Overall Risk Assessment (Hazard + Exposure + Sensitivity) – Capacity | | | | | 100% |

cumulative area of 92.67 km$^2$ followed by Southern District (8.796 km$^2$), Central and Western District (5.837 km$^2$), and Wan Chai District (4.589 km$^2$), as visible in the Figure 3.10 and tabulated in Table 3.4. The previous multi-hazard assessment study on the Central and Western District also agrees that the risk index is not very high. Southern District is categorized as the only district with an estimated area of 0.039 km$^2$ at a "very high" risk, indicating the robust resilience of Hong Kong Island to natural hazards.

Multi-hazard risk assessment is a comprehensive evaluation of the potentially imposing risks of intensifying hazards especially hydro-meteorological ones in the case of Hong Kong Island, Hong Kong. Through a comprehensive identification of the importance of four components: hazards, exposure, sensitivity, and capacity, extensive risk reduction action plans and resilience-building strategies can be formulated

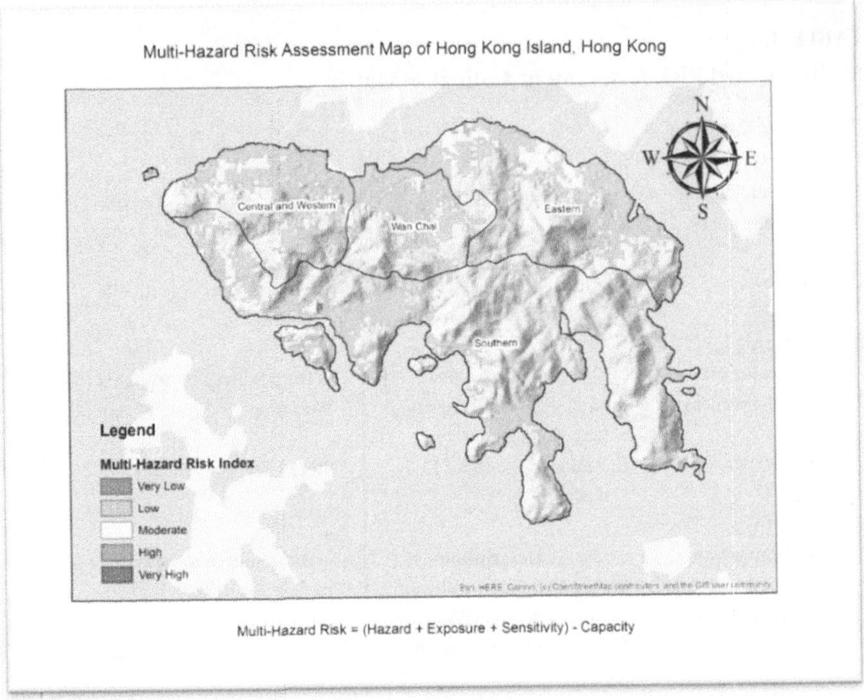


**FIGURE 3.10** Multi-hazard risk assessment map of Hong Kong Island.

as a means of promoting SDGs. Understanding the spatial contexts of potential risks, loss, and damage potential from natural hazards can be reduced through resilience planning and resource allocation. Moreover, major stakeholders and decision-makers can articulate the level of severity and probability of more common hazards in the region and prioritize the development of both required structured mitigation or adaptational measures, such as sea walls, nature-based solutions: mangrove forests, and unstructured disaster evacuation trainings and practices. Detailed multi-hazard risk assessment products such as maps can also be one of the preliminary disaster educational visualizing aids to promote a proactive approach and awareness and formulate community-based disaster risk reduction programs.

**TABLE 3.4**
**Multi-Hazard Risk Index Area in sq. km²**

| District Name | Very Low | Low | Medium | High | Very High | Percentage (%) |
|---|---|---|---|---|---|---|
| Central and Western | 0.018 | 5.425 | 0.394 | 0 | 0 | 5 |
| Wan Chai | 0 | 4.518 | 0.071 | 0 | 0 | 4 |
| Eastern | 0.078 | 62.3 | 14.419 | 15.869 | 0 | 83 |
| Southern | 0.012 | 8.177 | 0.515 | 0.0529 | 0.0392 | 8 |

Building socio-economic resilience using Geographic Information Systems in multi-hazard risk assessment can furthermore prevent the communities from poverty-disaster nexus, which in turn keeps SDG 1: no poverty on track. Moreover, resilience building through building structure modification, for sustainability purposes, promotes SDG 9: industry, innovation, and infrastructure, and SDG 11: sustainable cities and communities. Additionally, the knowledge products of underlying risks from multi-hazard risk assessment are crucial in the creation of adaptive sustainable development policies and practices with a conclusive foundation of local hazard profiles and drastically evolving risk landscapes.

## REFERENCES

Avtar, R., Aggarwal, R., Kharrazi, A., Kumar, P., & Agustiono, T. (n.d.). *Integrating Geospatial Information into the 2 implementation and monitoring of roadmaps for 3 achieving SDGs*.

Bera, S., Das, A., & Mazumder, T. (2022). Evaluation of machine learning, information theory and multi-criteria decision analysis methods for flood susceptibility mapping under varying spatial scale of analyses. *Remote Sensing Applications: Society and Environment, 25*. https://doi.org/10.1016/j.rsase.2021.100686

Bollin, C., Cárdenas, C., Hahn, H., & Vatsa, K. S. (n.d.). *Inter-American Development Bank, Regional Policy Dialogue Natural Disasters Network Disaster Risk Management by Communities and Local Governments*. https://publications.iadb.org/en/publications/english/viewer/Disaster-Risk-Management-by-Communities-and-Local-Governments.pdf

Danso, S. Y., Ma, Y., Osman, A., & Addo, I. Y. (2024). Integrating multi-criteria analysis and geospatial applications for mapping flood hazards in Sekondi-Takoradi Metropolis, Ghana. *Journal of African Earth Sciences, 209*. https://doi.org/10.1016/j.jafrearsci.2023.105102

Environment Bureau (2017). Hong Kong's Climate Action Plan 2030+. Environment Bureau.

ESRI. (2024). *Land Cover Explorer. ArcGIS Living Atlas of the World*. Retrieved November 12, 2022, from https://livingatlas.arcgis.com/landcoverexplorer/#mapCenter=-62.54228%2C-33.91918%2C11&timeExtent=2017%2C2024&year=2024

Hussain, M., Tayyab, M., Ullah, K., Ullah, S., Rahman, Z. U., Zhang, J., & Al-Shaibah, B. (2023). Development of a new integrated flood resilience model using machine learning with GIS-based multi-criteria decision analysis. *Urban Climate, 50*. https://doi.org/10.1016/j.uclim.2023.101589

Johnson, K., Depietri, Y., & Breil, M. (2016). Multi-hazard risk assessment of two Hong Kong districts. *International Journal of Disaster Risk Reduction*, *19*, 311–323. https://doi.org/10.1016/j.ijdrr.2016.08.023

Lai, Y., Li, J., Chen, Y. D., Chan, F. K. S., Gu, X., & Huang, S. (2023). Compound floods in Hong Kong: Hazards, triggers, and socio-economic consequences. *Journal of Hydrology: Regional Studies, 46*. https://doi.org/10.1016/j.ejrh.2023.101321

Lee, B. Y., Wong, W. T., Woo, W. C., & Kong Observatory, H. (2010). *Sea-level Rise and Storm Surge-Impacts of Climate Change on Hong Kong*. https://my.weather.gov.hk/tc/publica/reprint/files/r915.pdf

Sun, Y., Chau, P. H., Wong, M., & Woo, J. (2017). Place- and age-responsive disaster risk reduction for Hong Kong: Collaborative place audit and social vulnerability index for elders. *International Journal of Disaster Risk Science*, *8*(2), 121–133. https://doi.org/10.1007/s13753-017-0128-7

WorldPop. (2020). Global High Resolution Population Denominators Project - Funded by The Bill and Melinda Gates Foundation (OPP1134076). https://dx.doi.org/10.5258/SOTON/WP00675

# 4 Urban Disaster Planning and Flood Hazard Zonation from a Geomatic, GIS and Urban Planning Perspective

## *A Case Study of Sigatoka Town, Fiji Islands*

*Joeli Varo, William Feresi, Doni Wainoqolo, and Sujoy Kumar Jana*

## 4.1 INTRODUCTION

The Sigatoka flood hazard map was created and formulated through inter-governmental sector collaborative approach. The Ministry of Forestry collected the digital elevation model (DEM) through LiDar drone survey. They processed the data and handed over the pre-processed data for further analysis. Figure 4.1 illustrates the location of Sigatoka Town on Vitilevu Island. It covers 202.24 hectares of land mass with the growing urban population. There were no flood hazard assessment or demarcation has been done in Sigatoka Town which is a motivation for this research to provide a platform for future research. There were three research questions being asked to guide this research which were (i) what are the category of risks within the town boundary, (ii) how can geomatic assists in demarcating vulnerable zones and (iii) why is a need for this urgent research. Therefore, the main hypothesis of the research is to demarcate the hazard zone for Sigatoka Town and determine the flood zones accordingly.

## 4.2 METHODOLOGY

The methodology used for this analysis is known as the Analytical Hierarchical Process (AHP) using the multi-criteria analysis in the ArcGIS Pro and the excel sheet. This methodology has been used by Sekac et al. (2015) and Sekac et al. (2019) in Papua New Guinea and Varo et al. (2019, 2019b) in Fiji Islands. According to Shackleton (1936) revealed that the geology of Fiji Islands is very complex and fragile as well since it sits on the ring of fire. The analysis focus on the nine site-soil-geology


DOI: 10.1201/9781032683928-4

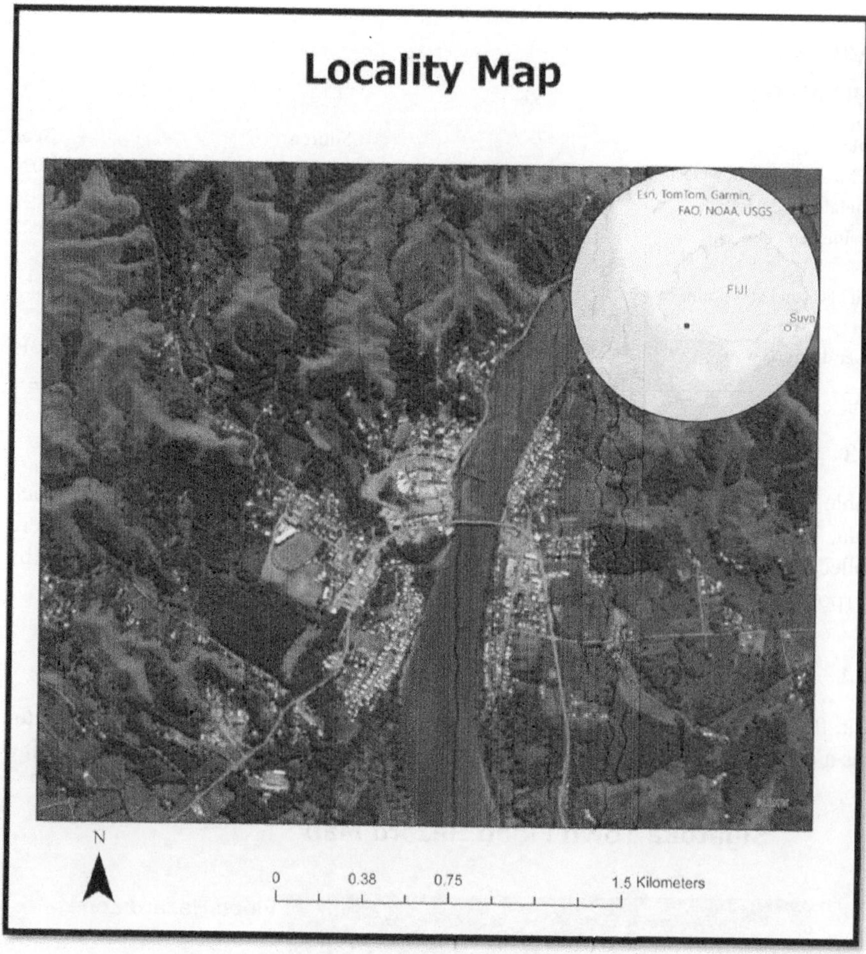


**FIGURE 4.1** Locality map of Sigatoka Town.

factors namely flow accumulation, river distance, elevation, land use, rainfall, slope, soil texture, soil drainage and geology. The formula shown below was used to derive the final flood hazard zone map of Sigatoka.

$$\text{FHI} = \sum^{n}(\text{RI})/\text{CI} \text{ Formular,}$$

where FHI = Flood Hazard Index

$\sum$ = sum

n = factor

RI = Random Index

CI = Consistency Index

**TABLE 4.1**
**Data Source**

| Data | Source | Year |
|---|---|---|
| DEM, slope and flow accumulation | Derived from drone – Ministry of Forestry | 2024 |
| Rainfall | Derived from Meteorological Department | 2023 |
| Geology and land use | Derived from Fiji Geological Map and Fiji Land use Capability Classification System | 2023 |
| Soil class and soil drainage | Derived from Fiji Land use Capability Classification System | 2023 |
| River distance | Derived from drone – Ministry of Forestry | 2014 |

## 4.3 DATA SOURCE

Table 4.1 showed the data used for this particular analysis and the sources of each data. This verifies and validates the accuracy of assessment since all the data were collected from the responsible government agencies and pre-processed to suit the purpose of this research.

## 4.4 RESULTS AND DISCUSSION

The flood hazard map as shown in Figure 4.2 revealed that Sigatoka Town center has a mixture of all the zones. Therefore, it reveals the tendency of flooding during

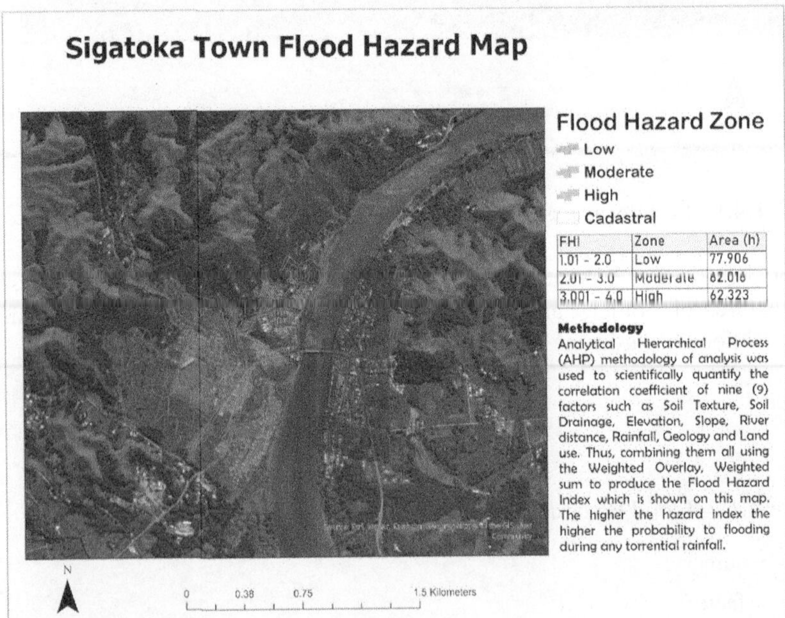


**FIGURE 4.2** Flood hazard map.

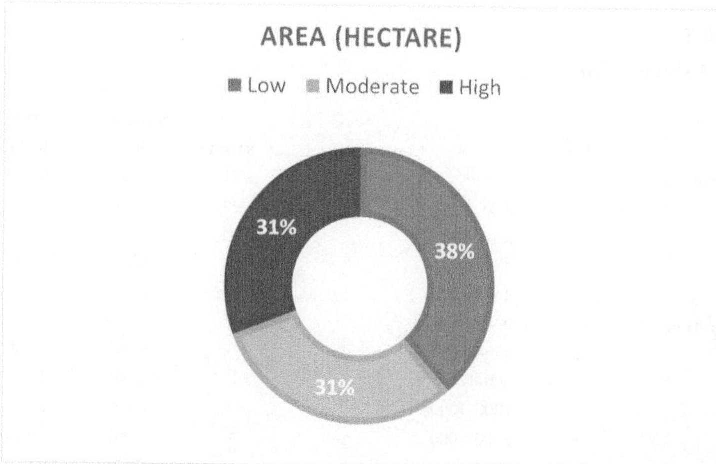


**FIGURE 4.3** Percentage of zone.

any torrential rainfall. The cadastral layer is overlaid onto this map to verify the lots present within which flood hazard zones.

The pie chart shown on Figure 4.3 shows the percentage coverage of each zone. Moderate and high zones percentage consist of 31% each while low zone consists of 38% of the total area coverage of 202.24 hectare of land.

Table 4.2 shown illustrated the percentage of area coverage for each zone. The Flood Hazard Index (FHI) illustrated the correlation coefficient of each zone. The FHI 1.01–1.0 represents low zone, 2.01–3.0 represents moderate zone and 3.01–4.0 represents high zone of flooding.

Table 4.3 illustrated the factors taken into consideration for the site-soil-geology analysis. It also reveals the individual weight of each factor, the classes of each factor, the ratings for each factor, normalize rate, area in hectare and the area percentage.

Figure 4.4a–4.4c illustrates the maps of each factor. The detailed area coverage, ranks and area percentage is detailed in Table 4.3, and Table 4.4 provides pair-wise comparison of the factor.

**TABLE 4.2**
**Percentage of Zone**

| FHI | Zone | Area (h) |
|---|---|---|
| 1.01–2.0 | Low | 77.906 |
| 2.01–3.0 | Moderate | 62.016 |
| 3.01–4.0 | High | 62.323 |

## TABLE 4.3
## Factors Information

| Factor | Weight | Classes | Ratings | Normalize Rate | Area (Hectare) | Area (%) |
|---|---|---|---|---|---|---|
| Flow accumulation | 0.296 | 0–2.53 | 1 | 0.062 | 93.7 | 37.7 |
| | | 2.53–3.16 | 2 | 0.098 | 77.3 | 31.1 |
| | | 3.16–3.92 | 3 | 0.161 | 52.7 | 21.2 |
| | | 3.92–4.91 | 4 | 0.262 | 20.0 | 8.0 |
| | | 4.91–6.29 | 5 | 0.416 | 4.7 | 1.9 |
| River dist. network (m) | 0.222 | 0–400 | 5 | 0.416 | 111.45 | 44.8 |
| | | 400–800 | 4 | 0.262 | 78.06 | 31.4 |
| | | 800–1200 | 3 | 0.161 | 47.85 | 19.2 |
| | | 1200–1600 | 2 | 0.098 | 11.26 | 4.5 |
| | | 1600–2000 | 1 | 0.062 | 0.00 | 0 |
| Elevation (m) | 0.157 | 58.464–61.482 | 5 | 0.416 | 46.19 | 18.6 |
| | | 61.483–67.089 | 4 | 0.262 | 61.02 | 24.5 |
| | | 67.09–70.539 | 3 | 0.161 | 44.44 | 17.9 |
| | | 70.54–83.478 | 2 | 0.098 | 54.04 | 21.7 |
| | | 83.479–168.44 | 1 | 0.062 | 42.94 | 17.3 |
| Land use | 0.11 | Mixed forest | 1 | 0.062 | 0 | 0 |
| | | Sparsely vegetated | 2 | 0.098 | 0 | 0 |
| | | Agricultural | 3 | 0.161 | 141.2 | 57.1 |
| | | Unimproved grassland | 4 | 0.262 | 11.6 | 4.7 |
| | | Urban wetland | 5 | 0.416 | 94.60 | 38.2 |
| Rainfall intensity (mm) | 0.077 | 219.60–254.14 | 1 | 0.062 | 0 | 0 |
| | | 254.14–278.32 | 2 | 0.098 | 248.64 | 100 |
| | | 278.32–303.36 | 3 | 0.161 | 0 | 0 |
| | | 303.36–350.86 | 4 | 0.262 | 0 | 0 |
| | | > 350.86 | 5 | 0.416 | 0 | 0 |
| Slope (degrees) | 0.053 | 0–1.325 | 5 | 0.416 | 43.5 | 17.5 |
| | | 1.326–3.977 | 4 | 0.262 | 54.1 | 21.8 |
| | | 3.978–9.944 | 3 | 0.161 | 53.0 | 21.3 |
| | | 9.945–21.213 | 2 | 0.099 | 48.8 | 19.7 |
| | | 21.214–84.524 | 1 | 0.062 | 48.8 | 19.7 |
| Soil class | 0.037 | Sand | 1 | 0.096 | 0 | 0 |
| | | Silt loam/loamy soil | 2 | 0.161 | 0 | 0 |
| | | Sandy clay loam | 3 | 0.277 | 248.33 | 100 |
| | | Peat | 4 | 0.466 | 0 | 0 |
| Soil drainage | 0.026 | Well drain | 1 | 0.096 | 0 | 0 |
| | | Imperfectly drain | 2 | 0.161 | 0.00 | 0 |
| | | Poorly drain | 3 | 0.277 | 247.46 | 100 |
| | | Water logged | 4 | 0.466 | 0.00 | 0 |
| Geology | 0.019 | Alluvium | 1 | 0.416 | 61.30 | 24.7 |
| | | Basic to intermediate volcanic | 2 | 0.262 | 45.43 | 18.3 |
| | | Coarse grain sedimentary | 3 | 0.161 | 0.015 | 0.0 |
| | | Mixed or undifferentiated sedimentary | 4 | 0.099 | 141.89 | 57.1 |
| | | Viti limestone | 5 | 0.062 | 0 | 0 |

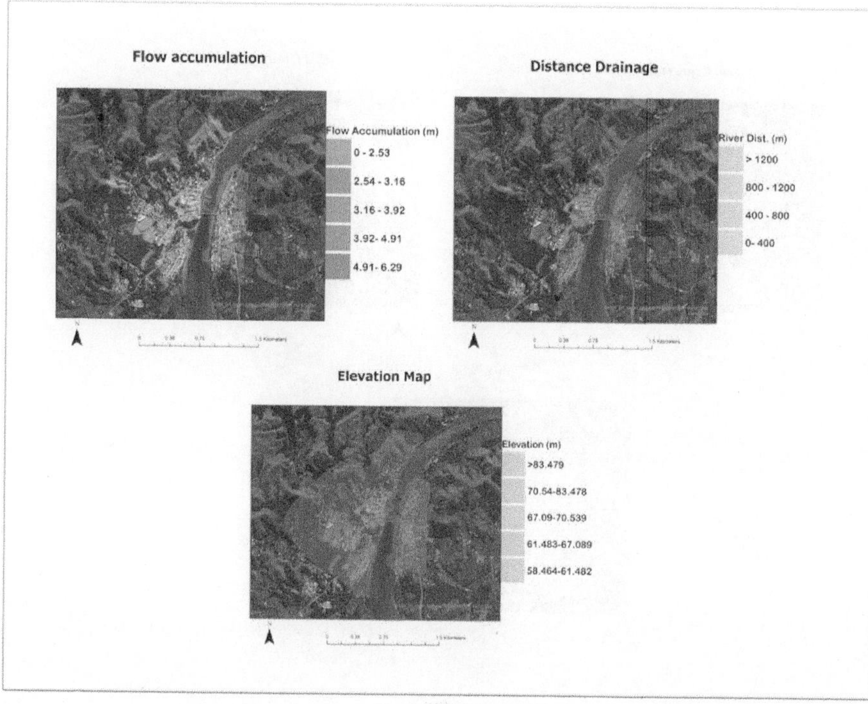


(A)

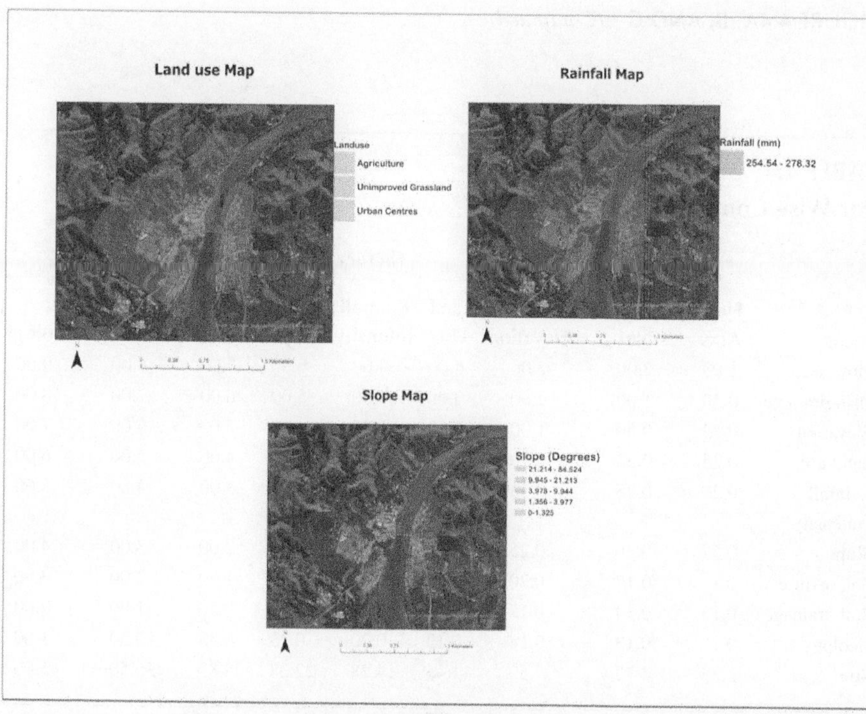


(B)

**FIGURE 4.4A, B, AND C** Factors mapping. *(Continued)*

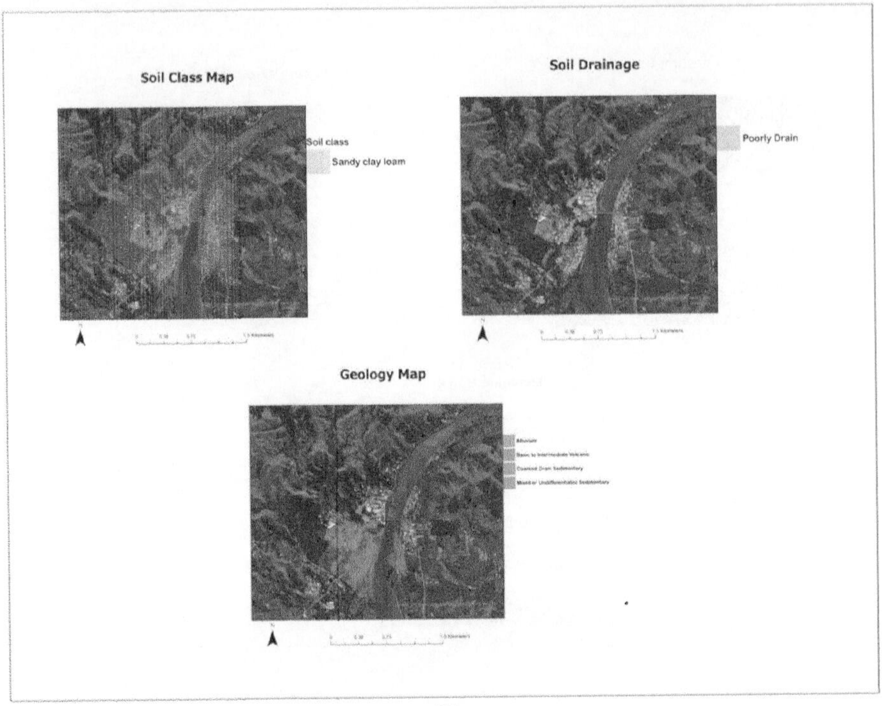


(C)

**FIGURE 4.4A, B, AND C** *(Continued)*

**TABLE 4.4**
**Par-Wise Comparison**

| | Factors | | | | | | | | |
|---|---|---|---|---|---|---|---|---|---|
| **Focus** | **Flow Acc.** | **Dist. Drain Net** | **Elevation** | **Land Use** | **Rainfall Intensity** | **Slope** | **Soil Texture** | **Soil Drainage** | **Geology** |
| Flow acc. | 1.00 | 2.00 | 3.00 | 4.00 | 3.00 | 6.00 | 7.00 | 8.00 | 9.00 |
| Dist. drain net | 0.50 | 1.00 | 2.00 | 3.00 | 4.00 | 5.00 | 6.00 | 7.00 | 8.00 |
| Elevation | 0.33 | 0.50 | 1.00 | 2.00 | 3.00 | 4.00 | 5.00 | 6.00 | 7.00 |
| Land use | 0.25 | 0.33 | 0.50 | 1.00 | 2.00 | 3.00 | 4.00 | 5.00 | 6.00 |
| Rainfall intensity | 0.20 | 0.25 | 0.33 | 0.50 | 1.00 | 2.00 | 3.00 | 4.00 | 5.00 |
| Slope | 0.17 | 0.20 | 0.25 | 0.33 | 0.50 | 1.00 | 2.00 | 3.00 | 4.00 |
| Soil texture | 0.14 | 0.17 | 0.20 | 0.25 | 0.33 | 0.50 | 1.00 | 2.00 | 3.00 |
| Soil drainage | 0.13 | 0.14 | 0.17 | 0.20 | 0.25 | 0.33 | 0.50 | 1.00 | 2.00 |
| Geology | 0.11 | 0.13 | 0.14 | 0.17 | 0.20 | 0.25 | 0.33 | 0.50 | 1.00 |
| **Sum** | 2.83 | 4.72 | 7.59 | 11.45 | 14.28 | 22.08 | 28.83 | 36.50 | 45.00 |

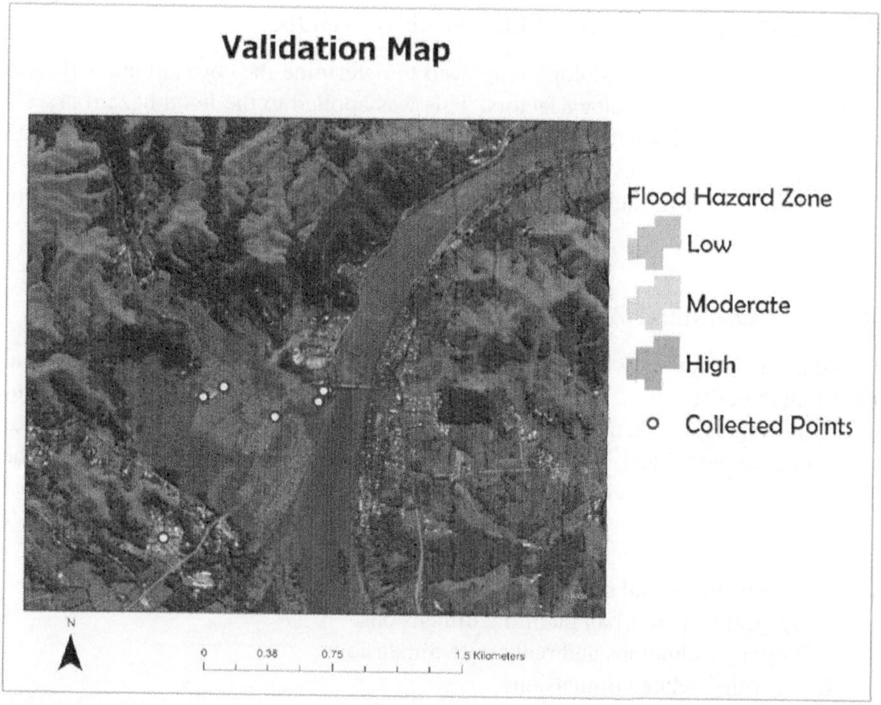


**FIGURE 4.5** Validation map.

## 4.5 VALIDATION OF RESULTS

Figure 4.5 illustrated the validation points collected on the ground on Thursday the 28 March 2024. There was a total of six points collected on the ground. Four points were on the high flood zone while only two points were recorded on the low flood zone. We also conduct a snap interview with community members regarding the location of the points and the risks of flooding. It was pleasing to notice that all of them corroborated with the validity of the points and its relationship top the risks of flooding during torrential rainfall.

## 4.6 IMPLICATIONS FOR ACADEMICIANS AND IMPLICATIONS FOR PRACTITIONERS

This research contributes significantly in academic and practitioners in terms of planning, budgeting, development and conservation of their land and future research. For academic, it provides a platform for future research and for practitioners to know where to build in the future to avoid socio-economic loses of their business and houses. This research also encourages the uses of artificial intelligence in data collection such as drone and python programming of large data sets for detailed analysis. Artificial intelligence has been used data collection, storing data and also analyzing the data to produce the desirable results presented in this research.

## 4.7 CONCLUSION AND RECOMMENDATION

AHP is a scientific methodology employed to determine the correlation coefficient and index of site-soil-geology factors. This was applied to the flood hazard assessment and the consistency ratio was well below the allowable inconsistencies and is acceptable. Thus, this flood hazard mapping is warranted to be implemented throughout our towns and cities in order to enhance planning and reduce disaster in the future year.

## 4.8 RECOMMENDATION

Flood hazard mapping is one of the components of disaster management and spatial planning in our societies. Since, this research will be developed into a national document, it is crucial to add multi-dimensional and integrated expert views on the socio-economic, geographical and human vulnerability assessments to this report. The following detail expert views are recommended to foster the wide-reaching impact of this report:

1. Human and social dimensions
2. Geography and urban planning dimensions
3. Property valuations and real estate dimensions
4. Any other related dimensions.

## REFERENCES

Sekac, T., Jana, S. K., and Pal, D. K., 2015, A Remote Sensing and GIS Approach to Assessing Multiple Factor Leading to Delineation of Flood Hazard Risk Zone in the Busu River Catchment, Morobe Province, Papua New Guinea. Melanesian Journal of Geomatics and Property Studies, ISSN (Online), Vol. 1, 2414–2557.

Sekac, T., Jana, S. K., Pal, I. and Pal, D. K., 2019, Application of Geospatial Technology in Earthquake Risk Assessment in Papua New Guinea. In International Expert Forum: Mainstreaming Resilience and Disaster Risk Reduction in Education. Springer Nature, Singapore. MRDRRE, 2017. Disaster Risk International Journal of Geoinformatics, Volume 16, No. 3, July–September 2020 Online ISSN 2673-0014/© Geoinformatics International 47 Reduction (Methods, Approaches and Practices). Springer, Singapore.185-218, 07/978-981-32-9527-8_12. DOI: 10.10.

Shackleton, R. M., 1936, Geology of VitiLevu. Geol. Surv. Fiji, Misc. Rep. 34. The Fiji Bureau of Statistic, 2017, Fiji Population Census by Province. https://www.statsfiji.gov.fj/fiji-census-2017-dashboard/

Varo, J., Sekac, T., Jana, S. K., & Pal, D. K., 2019a, Demarcation of Liquefaction Zones and Risk Reduction in Fiji Islands from a Geomatics Perspective: A Case Study of VitiLevu Island. Spatial Information Research, Vol. 27, 643658.

Varo, J., Sekac, T., & Jana, S. K., 2019b, Earthquake Hazard Micro Zonation in Vitilevu, International Journal of Recent Technology and Engineering. ISSN: 2277-3878, Vol. 8, Issue-12S, DOI: 10.35940/ijrte.B1257.0982S1119.

# 5 Learning Lessons from the Scenario Changes of Air Quality with Reference to the Pandemic in Delhi NCR

## *An MCDM Approach for Fighting Air Pollution*

*Rituparna Acharyya, Jyoti Prakash Hati, Niloy Pramanick, Sudipta Mukherjee, Suvojit Ganguly, Malini Singh, Tushar Ashok Tingote, Indrajit Pal, Sujoy Kumar Jana, Malay Pramanik, and Anirban Mukhopadhyay*

## 5.1 INTRODUCTION

The Central Pollution Control Board (CPCB) of India describes air quality as the volume of various pollutants in the atmosphere and their acceptable or permissible levels, which are commonly mentioned as standards (CPCB 2014). It indicates the quantity of minute suspended particles and other chemical compounds like oxides of carbon, sulfur dioxide, nitrogen dioxide, and ozone existent in the atmosphere. Air quality surpasses all extents within the atmosphere, ranging from local to global, providing feedback at each arena. It is responsible for several adverse impacts on health, the environment, climate, and monuments (Monks et al. 2009). The air quality index (AQI) is a tool to measure and publish information on air quality (e.g., good, satisfactory, poor) and its associated health impacts (CPCB 2014).

The intensity of air pollution is generally expressed as Air Quality (Monks et al. 2009). Jacobson, in Atmospheric Pollution: History, Science, and Regulation, defined pollution as a sufficiently high concentration of gases or anthropogenically emitted aerosol particles that cause direct or indirect damage to climate, ecosystems, and heritage (Jacobson 2002). Various air pollutants, such as carbon monoxide (CO), nitrogen oxides (NOx), sulfur dioxide (SO2), multiple forms of particulate matter (PM)

DOI: 10.1201/9781032683928-5

(SPM, $PM_{10}$, and $PM_{2.5}$), volatile organic compounds (VOCs), surface ozone ($O_3$), and heavy metals (cadmium, mercury), are responsible for adversely altering the atmospheric composition (Kampa and Castanas 2008). SPM (suspended particulate matter) is produced by chemical reactions in light with different gaseous substances existent in the atmosphere (Kuwata and Nishikawa 2005). According to the World Health Organization (WHO, 2016) in 2016, it was reported that 4.2 million premature deaths worldwide per year were caused by ambient (outdoor) air pollution in both urban and rural areas. Furthermore, around 58% of the deaths caused by ambient air pollution were caused by strokes and atherosclerotic heart disease. In comparison, 18% of fatalities were caused by severe conditions resulting from pulmonary obstruction such as chronic bronchitis and emphysema and acute infections in lower respiratory organs, and 6% of fatalities were caused by lung cancer (World Health Organization 2018).

Initially, the notion was that only industrialized and developed countries are prone to the ravages of air pollution. However, it has been noticed that even the middle-income and less-developed countries encounter the adverse consequences of air pollution. However, several regions worldwide have experienced changes in air quality due to lockdown measures implemented during the pandemic. Such changes may vary spatially due to variations in the discharged levels of pollutants, including vehicular emissions, fossil fuel combustion, and industrial production (Kerimray et al. 2020). A global instance of a few countries can be cited where a significant variation in air quality could be monitored because of precautionary measures to curb the pandemic, such as lockdowns due to a reduction in emissions from industries and vehicle power generation. In Almaty, Kazakhstan, during the lockdown period, the $PM_{2.5}$ concentration lowered by 21% (range: 6%–34%) in 2020 compared to the average $PM_{2.5}$ level, calculated on the same days in 2018–2019. However, the $PM_{2.5}$ concentration still exceeded the WHO standard. Also, the CO and $NO_2$ concentrations were substantially reduced, respectively, by 49% and 35% (Kerimray et al. 2020).

Similarly, NO2 concentration showed extensive reductions of approximately 53% in urban areas of Europe and 57% in Wuhan, China, when compared to the three-year average level (2017–2019); PM2.5 concentration had also reduced by 8% and 42% in urban regions of Europe and Wuhan, respectively (Ching and Kajino 2020). Also, results show that human activities in the Yangtze Delta Region in China lowered noticeably due to the pandemic; industrial and vehicular emissions were significantly reduced, which resulted in the lowering of $SO_2$, NOx, $PM_{2.5}$, and VOCs discharge by approximately 16%–26%, 29%–47%, 27%–46%, and 37%–57% during January to March 31, 2020 approximately (Li et al. 2020). The daily mean of $PM_{2.5}$ and $NO_2$ concentration within the New York City region between the first week of January and the first week of May in 2020 was reduced by approximately 36% and 51% (Zangari et al. 2020). However, worldwide studies demonstrate improvement in air quality due to reasons like social distancing and lockdown measures due to COVID-19. However, significant changes in NYC were not observed over 2015–2019, which might have occurred due to the analysis of both short- and long-term air quality variation (Zangari et al. 2020).

Like several countries, the influence of constricted human activities during the pandemic also had a positive consequence on the air quality of India. In comparison to previous years, all over India, $PM_{10}$, $PM_{2.5}$, CO, and $NO_2$ decreased around 43%, 31%, 10%, and 18%, respectively, possibly due to lockdown measures. However, there was a proportional increase in the level of $O_3$ by 17% (Sharma et al. 2020). In contrast to other mentioned pollutants, sulfur dioxide ($SO_2$) levels showed an increasing trend in Mumbai, Bengaluru, and Kolkata (Sathe et al. 2021). The AQI in north, south, west, east, and central India showed a decreasing trend of approximately 44%, 33%, 32%, 29%, and 15%, respectively (Sharma et al. 2020). The AQI lowered by 30% in 2020 compared to the preceding years; among the most polluted cities of India, Delhi observed the maximum positive impact in AQI with a reduction of almost 49%. During the pandemic, the eminent pollutant changed to $O_3$ in Gaya, Kolkata, Kanpur, and Nagpur, while, in Patna and Agra, $NO_2$ emerged as an imperious pollutant (Sharma et al. 2020).

The quality of air in different cities of India, pre-lockdown, however, portrays a different picture. The database was published by the WHO in April 2018 (WHO, 2018) and covered 100 countries worldwide between 2011 and 2016; a total of 14 cities out of the listed top 15 cities reported dangerous levels of $PM_{2.5}$ pollution from India. According to a WHO report released in 2018 among various megacities worldwide, Delhi occupies the apex position in the list of $PM_{10}$ pollution (Guttikunda et al. 2019). Delhi, Agra, and a few other cities like Mumbai, Kolkata, Chennai, and Hyderabad were repeatedly covered in various research studies conducted over the years since 2000 to identify the potential sources of $PM_{2.5}$ and $PM_{10}$ (Guttikunda et al. 2019). Delhi has long been recognized as among the most polluted cities in the world. The Delhi government announced a health emergency in November 2019 when the city registered a $PM_{2.5}$ level of 625 micrograms per cubic meter ($\mu g/m^3$) of air, which was 24 times the WHO's recommended standards (Bharadwaj 2020). The volume of PM is higher in Agra than in other urban regions in India (Gupta et al. 2017). Delhi's air pollution issue may well have contributed to increasing the COVID-19 devastation throughout the city. Long-term $NO_2$ and $PM_{2.5}$ exposure can influence COVID-19 fatality (Bharadwaj 2020). The PM could sustain the SARS-CoV-2 pathogen, which induces COVID-19, allowing it to persist in the atmosphere and proliferate. The $SO_2$ level of both the cities, Delhi and Agra, was at par with the annual standard of PCB, which accounts for 50 $\mu g/m^3$, while in the case of $NO_2$, the level exceeded the annual standard of 40 $\mu g/m^3$ in both the cities (Guttikunda et al. 2019).

These are some major contributing factors to the reduction in pollution in Delhi National Capital Region (NCR): a 97% decrease in general traffic and a 91% reduction in lorries and commercial vehicles reaching the capital all through April, particularly in comparison to the pre-lockdown period of December–January, according to a report by the "Centre for Science and Environment" (PTI 2020). The AQI stayed in the "moderate" range between May 18 and June 5. The monsoon season kept pollution conditions under control from July through September. August has been the cleanest month of the year for air quality, with a mean AQI of 63.8. In August, there were 4 "good" air dates recorded, the most in any period until 2015 since the CPCB began measuring AQI (PTI 2020).

In light of the findings, as mentioned earlier, the current study aims to determine the influence of lockdown on air pollution levels in Delhi NCR (National Capital Region) and its adjacent regions compared to previous years. This study attempts to address air pollution levels in terms of $PM_{10}$, $PM_{2.5}$, $NO_2$, and $SO_2$ and their distributions in sensitive regions of Delhi NCR and its neighboring districts. A better understanding of the region's pollution dynamics and public health risks will be documented in this study. Given the environmental consequences of the lockdown, an effort has been undertaken to estimate the spatio-temporal variations in the amount of air pollution in India over the post-lockdown era compared to the preceding era.

The objectives of this research are to evaluate and compare the extent of atmospheric pollutants in the country study area before, during, and then after the lockdown, including the concentrations of air pollutants ($PM_{10}$, $PM_{2.5}$, $NO_2$, and $SO_2$), and to investigate the potency of the lockdown in enhancing environmental quality. Aside from that, the estimated AQI has been estimated for the current study by following NAQI (National Air Quality Index) by CPCB. It is calculated by combining a weighted mean of the criterion pollutants, including $PM_{10}$, $PM_{2.5}$, $NO_2$, and $SO_2$ into a uniform value matrix over the same study period of April before (2019), during (2020), and after the complete lockdown (2021), to investigate the lockdown's effectiveness in improving environmental quality. This study also tried to propose a framework for effective management of air quality standards in Delhi and the NCR region. This framework will help the management bodies and policymakers such as the government as well as stakeholders to tackle the problem in an informed manner.

## 5.2 STUDY AREA

The current study has been undertaken across the Delhi NCR, encompassing five vulnerable urban regions based on various site characteristics, such as commercial, industrial, institutional, and residential. The current study area (Figure 5.1) includes the Delhi NCR, including surrounding districts of Kaithal and Hisar in Haryana; Bijnor, Haridwar, Saharanpur, Aligarh, Etah, Hathras, Firozabad, Mathura, Bharatpur, Firozabad, and Agra in Uttar Pradesh; shown in Figure 5.1. NCR is located between 27.60°N and 29.30°N and 76.20°E and 78.40°E, with a dense population (790 people/sq. km) encompassing 4 states—Delhi, Haryana, Uttar Pradesh, and Rajasthan—and with a combination of 24 four districts. The NCR now spans an area of around 58,332 sq. km, with Haryana having the largest area (28,545 sq. km), trailed by Uttar Pradesh (14,858 sq. km), Rajasthan (13,446 sq. km), and then Delhi (1,483 sq. km) (Hazarika et al. 2019). The capital (Delhi) does have a semi-arid environment and is physically situated between 28.39°N to 28.88°N and 76.82°E to 77.34°E. It is bounded to the north by the Himalayan Mountain range, to the south by the central heated peninsular area, to the northeast by mountainous terrain, and to the west by the Great Indian Thar Desert (Yadav et al. 2016). Delhi's climate is distinguished by a hot and dry summer and chilly winter (Khillare et al. 2008). The following major national highways (NH) go through Delhi: NH-1, -2, -8, -10, -24, and -48. In the summer, dust carried by winds (South–South west) from the Thar Desert influences Delhi's climate. Winter is distinguished by temperature reversal, contributing to increased aerosol concentration by dropping the planetary boundary

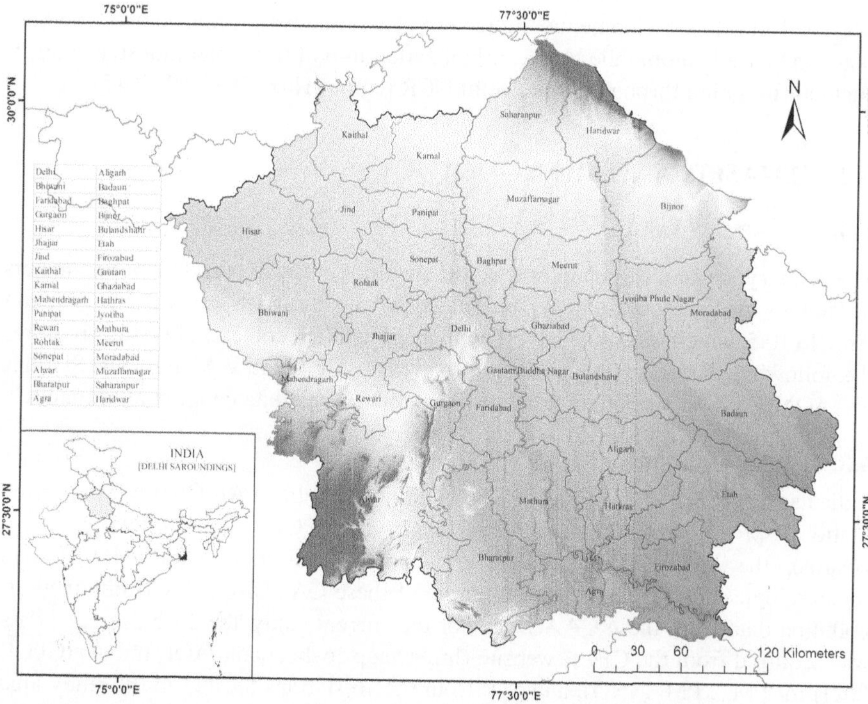


**FIGURE 5.1** Location map of the study area selected for the current study.

zone (Yadav et al. 2014). The average speed of wind in the capital has been estimated to be around 1.0 and 2.0 m/s (Srivastava & Jain, 2007; Srivastava et al., 2009). The district of Agra examined in this study has a long history and is rich in historical sites (Directorate of Census Operations Uttar Pradesh 2011). Nevertheless, Agra's thriving small-scale businesses (leather, handicrafts, iron factories), tourism sector, electronic parts manufacture, thermal power plants, brick industry, and agriculture constitute the basis of the city's economy (Guttikunda et al. 2019). The exquisite architecture of the Taj Mahal, sculpted from white marble stone, is being seriously harmed as pollution levels increase (Guttikunda et al. 2019). To preserve the inheritance, the Supreme Court declared a 50-km safety zone around the Taj Mahal in 2000, termed the TTZ (Taj Trapezium Zone), which should be free of industrial and traffic pollutants. Nevertheless, this strategy proved ineffective since the impact was insignificant, resulting in an exponential increase in $PM_{2.5}$ emissions in Agra, which eventually ranked the city as the fourth most polluted city in India in 2016 (Guttikunda et al. 2019; Kumar et al. 2020). The NCRPB (National Capital Region Planning Board under the Ministry of Urban Growth, Government of India Act, 1985) established the NCR intending to encourage the region's well-coordinated development. The NCRTC (National Capital Region Transport Corporation), established in 2013, is responsible for implementing the RRTS (Regional Rapid Transit System) project across the NCR to ensure the equitable and sustainable growth of metropolitan regions. In terms of air pollution, the capital region's air quality has worsened due to developmental projects, vehicle intensities, and industrial emissions. Increased automobile traffic, urbanization in buildings, and industry contribute to air pollution throughout the Delhi NCR region (Hazarika et al. 2017).

## 5.3 DATASETS AND METHODOLOGY

### 5.3.1 Datasets Used

The primary focus of this chapter is on the specifics of the data and input parameters needed for the present research. The procedures were carried out with the program ArcGIS 10.5, which was used to work with maps in a GIS context. This study requires the following datasets: (i) Continuous Ambient Air Quality Monitoring Stations (CAAQMS)-based air pollutant data; (ii) satellite-based data on air contaminants.

#### 5.3.1.1 Air Pollutants

India has a network of 231 CAAQMS (Continuous Ambient Air Quality Monitoring Stations) spread across the nation. The CPCB and SPCBs (State Pollution Control Boards), the DPCC (Delhi Pollution Control Committee), and IMD (Indian Meteorological Department) are in charge of these CAAQMS. CPCB obtained air pollution data from these CAAQMS. For the current study, the 24-h averaged data was acquired from the CPCB website (https://app.cpcbccr.com/AQI_India/) (CPCB 2021) for $PM_{10}$, $PM_{2.5}$, $NO_2$, and $SO_2$ from the 16 stations located in the study area for April 2019–2021. The data from April 2019 was only used to compare temporal alterations in the concentration dimensions of air contaminants for 2019 and 2020. In contrast, data from April 2021 was incorporated into statistical assessment for comparative analysis of intensity concentrations of air pollutants observed before, during, and after the nationwide lockdown.

#### 5.3.1.2 Satellite Data Analysis

The satellite imageries depicting concentrations of $NO_2$ and $SO_2$ (spatial resolution: 0.25°-level-3 daily OMI tropospheric) for April 2019–2021 were acquired from the Giovanni interface (https://giovanni.gsfc.nasa.gov/giovanni/). Then the imageries illustrating the concentration of PM2.5 and PM10 derived from the European Copernicus Atmosphere Monitoring Service (CAMS) (https://atmosphere.copernicus.eu/) (ECMWF 2021) at a spatial resolution of 1°, depicted in Table 5.1.

### 5.3.2 Methodology

#### 5.3.2.1 Air Quality Index Estimation

The AQI was calculated using data from air pollutants as input parameters, as shown in Figure 5.2.

Finally, the study area's spatio-temporal fluctuation of air pollutants ($PM_{10}$, $PM_{2.5}$, $NO_2$, and $SO_2$) was assessed before, during, and after the lockdown period (2019–2021). The CPCB released real-time data on NAQI across 231 CAAQMS (https://app.cpcbccr.com/AQI India/). NAQI is determined by combining a weighted mean of the criterion pollutants, including $PM_{10}$, $PM_{2.5}$, $NO_2$, $SO_2$, $O_3$, CO, $NH_3$, and Pb, into

**TABLE 5.1**
**Datasets Used for the Current Study**

| Data Type | Description | Spatial Resolution | Locations (CPCB, SPCBS) | Time Interval | Data Sources |
|---|---|---|---|---|---|
| $PM_{10}$ | Particulate matter (PM) (<10 μm diameter) | 1° | **Delhi:** Ashok Vihar **Haryana:** H.B. Colony, Bhiwani Vikas Sadan, Gurugram Urban Estate-II, Hisar Police Lines, Jind Rishi Nagar, Kaithal Sector-12, Karnal Sector-18, Panipat Murthal, Sonipat **Rajasthan:** Moti Doongri, Alwar **Uttar Pradesh:** Yamunapuram, Bulandshahr Sector – 125, Noida, Sanjay Nagar, Ghaziabad, Lajpat Nagar, Moradabad, New Mandi, Muzaffarnagar | April (2019–2021) | https://app.cpcbccr.com/AQI_India/ |
| $PM_{2.5}$ | PM (<2.5 μm diameter) | 1° | | | https://giovanni.gsfc.nasa.gov/giovanni/ |
| $NO_2$ | Nitrogen dioxide (μg/m³) | 0.25° | | | https://cpcb.nic.in/real-time-air-qulity-data/ |
| $SO_2$ | Sulfur dioxide (μg/m³) | 0.25° | | | https://app.cpcbccr.com/AQI_India/ |
| $CO_2$ | Carbon dioxide (PPM) | 2° * 0.25° | | | |

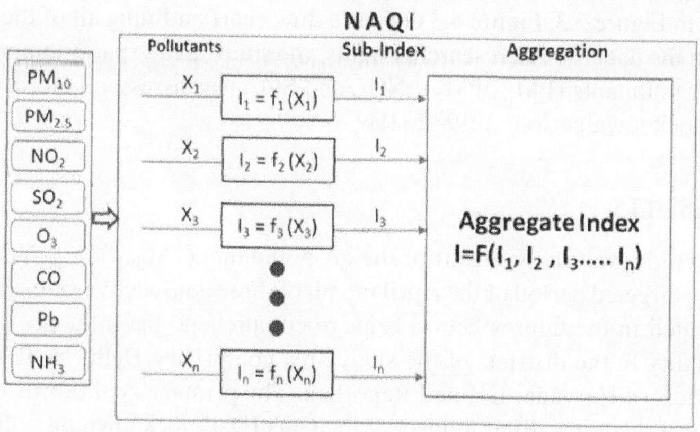


**FIGURE 5.2** An illustration with reference to CPCB's systematic method (CPCB 2014) for calculating the NAQI from observations of the major pollutants.

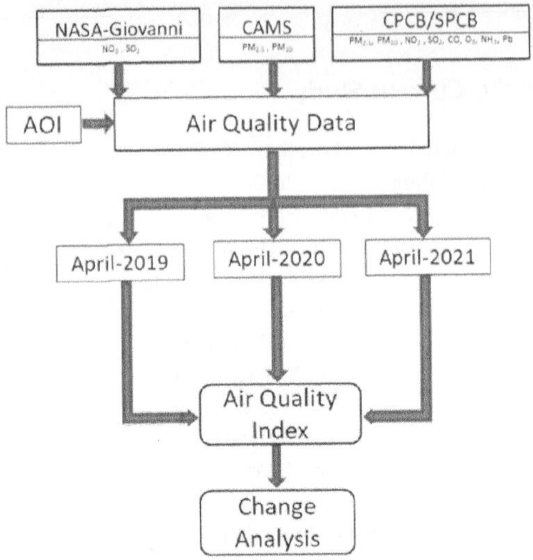


**FIGURE 5.3** Flow chart showing the methodology of the current study.

a uniform value matrix. In a nutshell, the NAQI was computed in two stages. First, each criterion pollutant's sub-index was computed. Second, all sub-indices are subsequently combined to form a single index. The Document on NAQI discusses the formulae used to calculate the NAQI in detail (CPCB 2014). The daily mean NAQI data from the CAAQMS were obtained from the CPCB site (https://app.cpcbccr.com/AQI_India/) (CPCB 2021) for April during 2019–2021. If a location has more than one CAAQMS, one daily mean NAQI value from across all CAAQMS has also been observed for that location. Furthermore, not all the CAAQMS are operational at all times.

The AQI was then calculated using data from air pollutants as input parameters, as shown in Figure 5.3. Figure 5.3 depicts a flow chart outlining all of the procedures utilized in the data for this research. Finally, the study area's spatio-temporal fluctuation of air pollutants ($PM_{10}$, $PM_{2.5}$, $NO_2$, and $SO_2$) was assessed before, during, and after the lockdown period (2019–2021).

## 5.4 RESULTS

The observed level (daily mean) of the air pollutants ($PM_{10}$, $PM_{2.5}$, $NO_2$, and $SO_2$) during the targeted period of the April month of three consecutive years (2019–2021), demonstrated in the figures below, helps to comprehend the influence of lockdown on air quality in the districts of the study area comprising Delhi NCR and its adjacent districts of Haryana, UP, and Rajasthan. The primary goal of this chapter is to determine if there is a direct impact of the COVID-19 lockdown on pollution levels via the findings. To determine this direct impact of lockdown, the spatio-temporal changes of pollutant levels before, during, and after complete lockdown have been evaluated, using the April months of 2019–2021.

### 5.4.1 Spatio-Temporal Variations of Air Pollutants

#### 5.4.1.1 PM2.5 (Particulate Matter < 2.5 μm)

According to USEPA, $PM_{2.5}$ includes fine particles inhaled with diameters of 2.5 μm and smaller. $PM_{2.5}$ as a pollutant came into consideration in 2009 and was first measured in 2016 at the manual stations in 2016 (Guttikunda et al. 2019). The utmost standard set for $PM_{2.5}$ in ambient air for a 24-h mean value is 25 and 60 μg/m$^3$, as mentioned by World Health Organization, Occupational and Environmental Health Team (2006) and CPCB (2009), respectively (Garg et al. 2020). However, $PM_{2.5}$ has more impact, even when it has a low concentration in the atmosphere, and it is often linked to several hazardous impacts on health compared to other pollutants (Saxena and Raj 2021). Industrial and vehicular exhausts, domestic cooking and heating, open burning of waste, diesel generator sets, and construction dust are the prominent sources of atmospheric $PM_{2.5}$ pollution (Guttikunda et al. 2019). Figure 5.4 shows images illustrating the concentration of $PM_{2.5}$ derived from CAMS (https://atmosphere.copernicus.eu/) of April 2019–2021, depicting the pre-lockdown, complete lockdown, and post-lockdown or partial lockdown period. It can be observed from the images in Figure 5.4 that the level of $PM_{2.5}$ was very high during April 2019, which drastically dropped during the complete lockdown in April 2020 and again rose to a moderate level in April 2021.

As per CPCB data (https://cpcb.nic.in/real-time-air-qulity-data/), the mean $PM_{2.5}$ level, which was 79.27 μg/m$^3$ during April 2019, decreased to 40.28 μg/m$^3$ in April 2020 and has again increased to 68.90 μg/m$^3$ in April 2021. The maximum $PM_{2.5}$ level was recorded in Ghaziabad (Sanjay Nagar – UPPCB) of Uttar Pradesh was 104.70 μg/m$^3$ in April 2019; the maximum level was reduced to 59.54 μg/m$^3$ in April 2020. However, it has again increased to 101.29 μg/m$^3$ in April 2021. The minimum $PM_{2.5}$ level recorded in Panipat (Sector-18, HSPCB) of Haryana was 39.17 μg/m$^3$ in April 2019. In April 2020, the minimum $PM_{2.5}$ level (19 μg/m$^3$) was recorded in Sanjay Nagar (Ghaziabad – UPPCB) of Uttar Pradesh, which was comparatively lower than that of 2019. In April 2021, the lowest concentration of $PM_{2.5}$ again slightly increased to 21 μg/m$^3$ recorded in Haryana's Bhiwani (H.B. Colony – HSPCB). These observations prove that the complete lockdown had a profound effect on reducing pollutant $PM_{2.5}$ in ambient air over the study area.

#### 5.4.1.2 $PM_{10}$ (Particulate Matter <10 μm)

The $PM_{10}$ consists of those PMs whose aerodynamic diameter is lower than 10 μm. They have the highest human lung penetration-affinity, resulting in increased respiratory and cardiac illnesses and, in extreme cases, even premature demise (Guttikunda 2012). In ambient or outdoor air, the maximum limit of the 24-h average of PM10 as recommended by WHO (2006) is 50 μg/m$^3$ and the permissible limit at 100 μg/m$^3$ (Garg et al. 2020). Since 2005 the volume of PM10 has consistently increased and ultimately accounted for a rise more significant than 2.5 times (Guttikunda 2012). In 2010, the yearly mean outdoor $PM_{10}$ concentrations accounted for 260 μg/m$^3$, almost four times more than the national standard and 13 times greater than the guidelines mentioned by the WHO (Guttikunda 2012). Major sources of $PM_{10}$ consist of the combustion of coal, kerosene, fossil fuels, biomass, and dust (Guttikunda et al. 2019). Figure 5.5 depicts the concentration of $PM_{10}$ obtained from CAMS

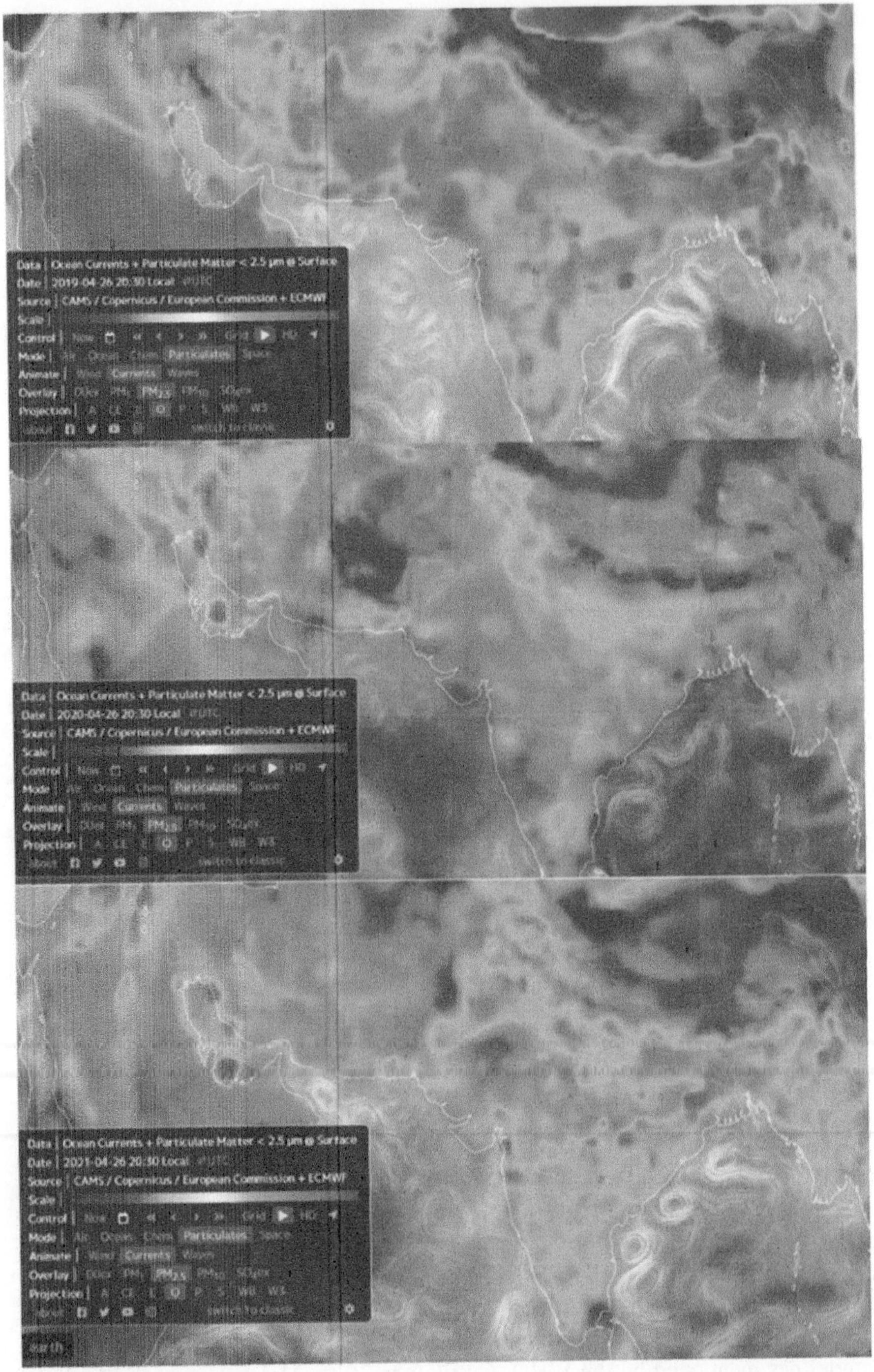


**FIGURE 5.4** Variation in concentration during April 2019–2021 of $PM_{2.5}$ ($\mu g/m^3$).

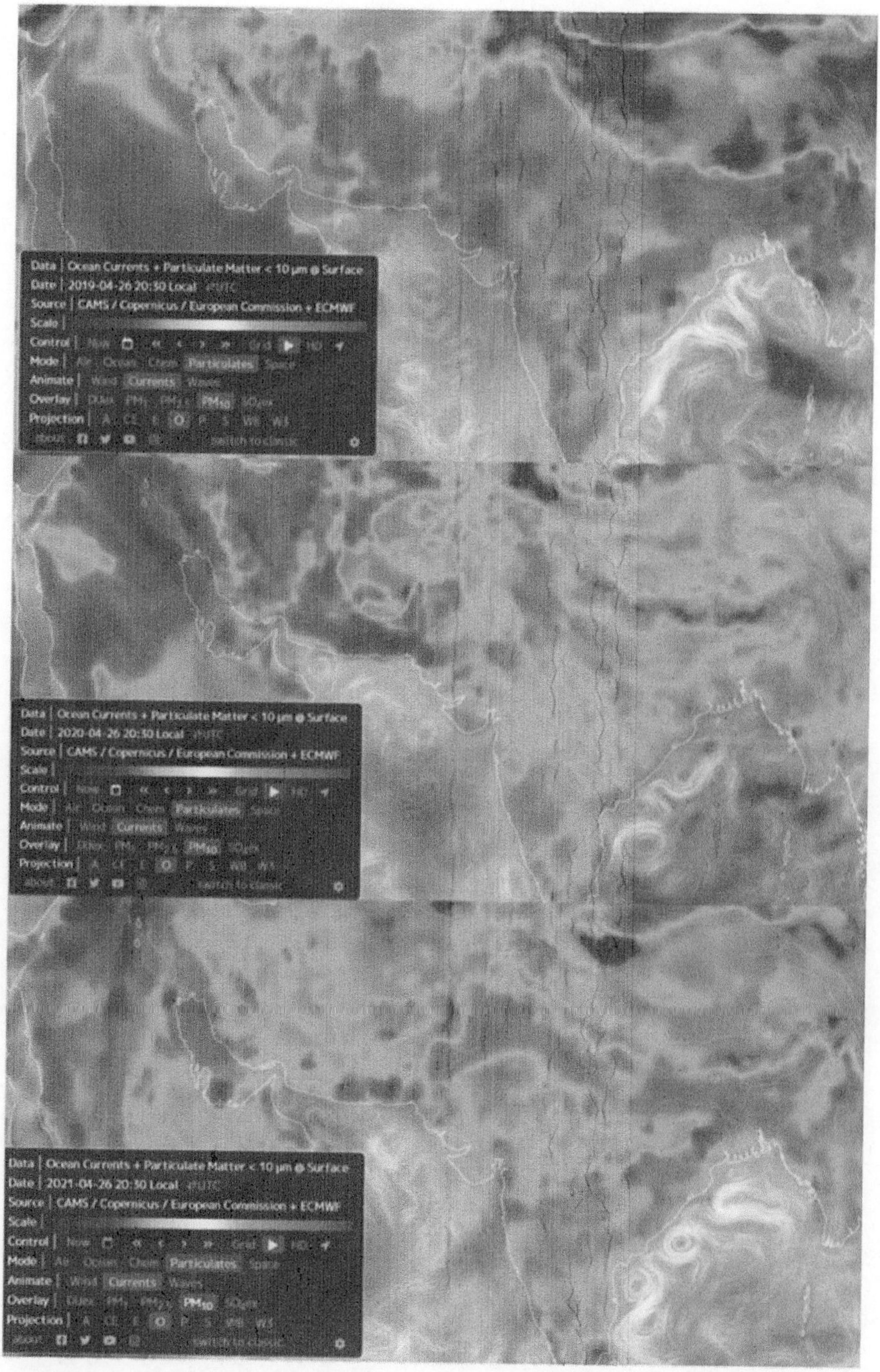


**FIGURE 5.5** Variation in concentration from April 2019 to 2021 of PM10 (µg/m³).

(https://atmosphere.copernicus.eu/) for April 2019–2021, showing the pre-lockdown, total lockdown, and post-lockdown or partial lockdown periods, similar to the $PM_{2.5}$ levels. According to the images in Figure 5.5, the level of $PM_{10}$ was very high in April 2019 and then decreased dramatically during the full lockdown in April 2020 before rising to a reasonable level in April 2021. According to CPCB statistics and Figure 5.5, the mean $PM_{10}$ level in April 2019 was 198.50 $\mu g/m^3$, dropped to 107.7 $\mu g/m^3$ in April 2020, and rose to 193.63 $\mu g/m^3$ in April 2021. In April of this year, the highest $PM_{10}$ level in Ghaziabad (Sanjay Nagar – UPPCB) in Uttar Pradesh was 303.64 $\mu g/m^3$. The highest $PM_{10}$ level in Bhiwani (H.B. Colony) was 188.06 $g/m^3$ in April 2020 and then rose to 188.06 $\mu g/m^3$ in April 2021. Though there has been a reduction in concentrations of $PM_{10}$ during the lockdown in the regions mentioned above, throughout the whole period, the $PM_{10}$ level exceeds the standard set by WHO (2006), making the air quality harmful.

The lowest $PM_{10}$ level in Panipat (Sector-18 – HSPCB) in Haryana was 39.17 $\mu g/m^3$ in April 2019. In Moradabad (Lajpat Nagar – UPPCB) in Uttar Pradesh, the minimum $PM_{10}$ level (34.43 $\mu g/m^3$) was reported in April 2020, lower than that of 2019. These levels of $PM_{10}$ are within the standards of WHO (2006). In April 2021, the lowest concentration of $PM_{10}$ in Alwar (Moti Doongri – RSPCB) in Rajasthan rose significantly to 84.11 $\mu g/m^3$. These findings show that the full lockdown had a significant impact on lowering pollutant $PM_{10}$ levels in the ambient air across the study region.

#### 5.4.1.3 SO2 (Sulfur Dioxide)

$SO_2$ as gas is colorless along with a sharp odor (WHO, 2018). The main origin of $SO_2$ is fossil fuel (coal and oil) incineration and the smelting of sulfur-rich mineral ores. The anthropogenic sources comprise fossil fuel combustion for domestic purposes, power generation, and transport (WHO, 2018). The standard set for $SO_2$ for a mean of 24 h by CPCB (2009) is 80 $mg/m^3$, whereas the standard set by WHO (2006) is 40 $mg/m^3$ in ambient air (2006) (Garg et al. 2020). $SO_2$ mainly causes respiratory diseases, affects the lungs, and is responsible for irritation in the eyes (WHO, 2018). In combination with water, $SO_2$ forms sulfuric acid, which forms the main constituent of acid rain, a major cause of deforestation (WHO, 2018). Figure 5.6 shows SO2 concentrations across the study area acquired from NASA-Giovanni (https://giovanni.gsfc.nasa.gov/giovanni/) for April 2019–2021, indicating the pre-lockdown, complete lockdown, and post-lockdown or partial lockdown periods, comparable to $SO_2$ levels. According to the images in Figure 5.6, the level of $SO_2$ was very high in April 2019. $SO_2$ levels significantly reduced during the complete lockdown in April 2020, improving the air quality, before rising to a moderate level in April 2021.

According to CPCB data (https://cpcb.nic.in/real-time-air-qulity-data/) in support of Figure 5.6, the average $SO_2$ level in April 2019 was 25.63 $\mu g/m^3$, dropped to 16.46 $\mu g/m^3$ in April 2020, and climbed to 24.25 $\mu g/m^3$ in April 2021. In April 2020, the highest $SO_2$ level at Bulandshahr (Yamunapuram – UPPCB) in Uttar Pradesh was 46.87 $\mu g/m^3$. In April 2020, the highest $SO_2$ level in Bhiwani (H.B. Colony) was 31.73 $\mu g/m^3$, which is the satisfactory standard set by WHO (2006). The maximum $SO_2$ level was 49.78 $g/m^3$ in April 2021, which exceeds WHO's standard. The minimum $SO_2$ level was 9.74 $\mu g/m^3$ in April 2019 in Kaithal (Rishi Nagar – HSPCB) in

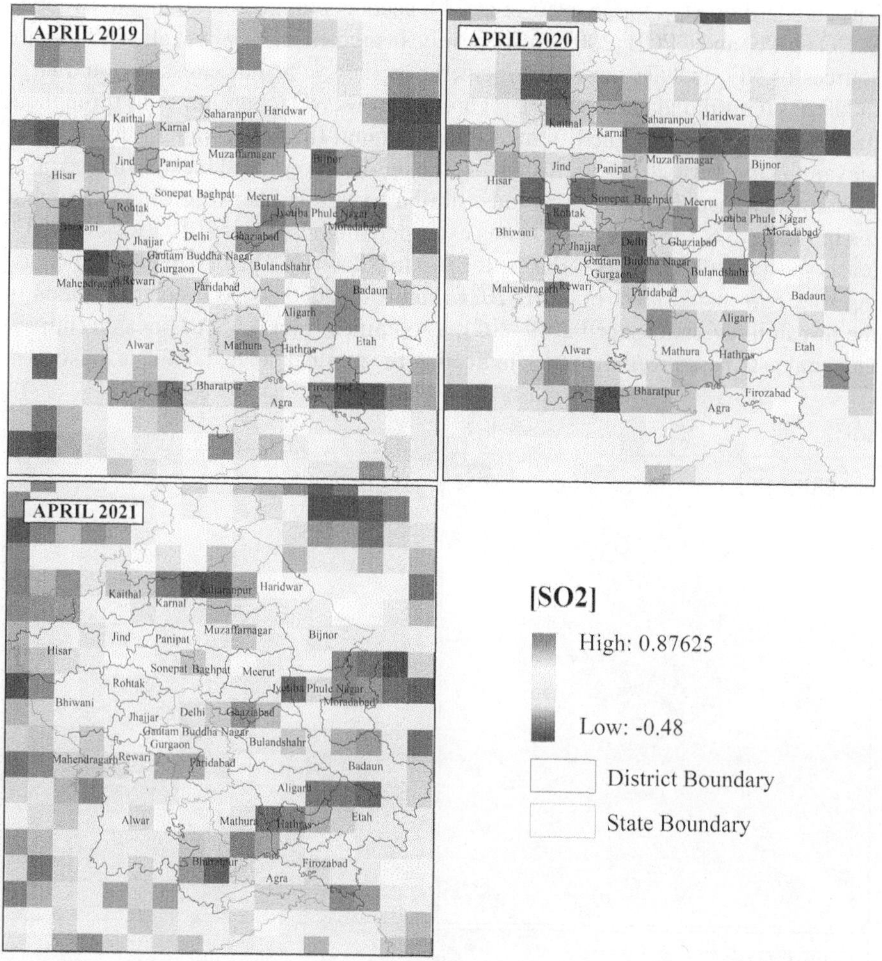


FIGURE 5.6 Variation in concentration from April 2019 to 2021 of SO2 (μg/m³).

Haryana. In Gurugram (Vikas Sadan – HSPCB), the minimum $SO_2$ level (6.11 μg/m³) was observed in April 2020, lower than in 2019. In April 2021, the lowest $SO_2$ concentration in Jind (Police Lines – HSPCB) in Haryana increased to 12.91 g/m³, higher than in 2019. The figures mentioned above of minimum $SO_2$ levels are within the standards set by WHO (2006). Thus, like those of $PM_{2.5}$, and $PM_{10}$, this data reveals that the full lockdown had a significant influence on lowering pollutant $SO_2$ levels in the ambient air and improving its quality across the study area.

#### 5.4.1.4 NO2 (Nitrogen Dioxide)

NO2 is an oxidant gas that is primarily produced due to fossil fuel combustion. It is responsible for triggering asthma (Shima and Adachi 2000). Nitrate aerosols are products of NO2, and they constitute an important fraction of $PM_{2.5}$; in the existence of ultraviolet light, it becomes a major component of ozone (WHO, 2018).

The standard of $NO_2$ for an average of 24 h accounts for 20 and 80 mg/m$^3$ as set by WHO (2006) and CPCB (2009), respectively, in ambient air (Garg et al. 2020). Major sources of $NO_2$ include fossil fuel incineration (i.e., vehicular emissions) and large industries (Guttikunda et al. 2019). Figure 5.7 shows $NO_2$ concentrations throughout the research region obtained from NASA-Giovanni (https://giovanni.gsfc.nasa.gov/giovanni/) for April 2019–2021, showing the pre-lockdown, full lockdown, and post-lockdown or partial lockdown periods, similar to $NO_2$ levels. According to the photos in Figure 5.6, the level of $NO_2$ was extremely high in April 2019. However, $NO_2$ levels substantially decreased during the full lockdown in April 2020 before increasing to a reasonable level in April 2021. Figure 5.7 depicts $NO_2$ concentrations in the research region for April 2019–2021, as obtained from NASA-Giovanni (https://giovanni.gsfc.nasa.gov/giovanni/), indicating the pre-lockdown, complete lockdown,

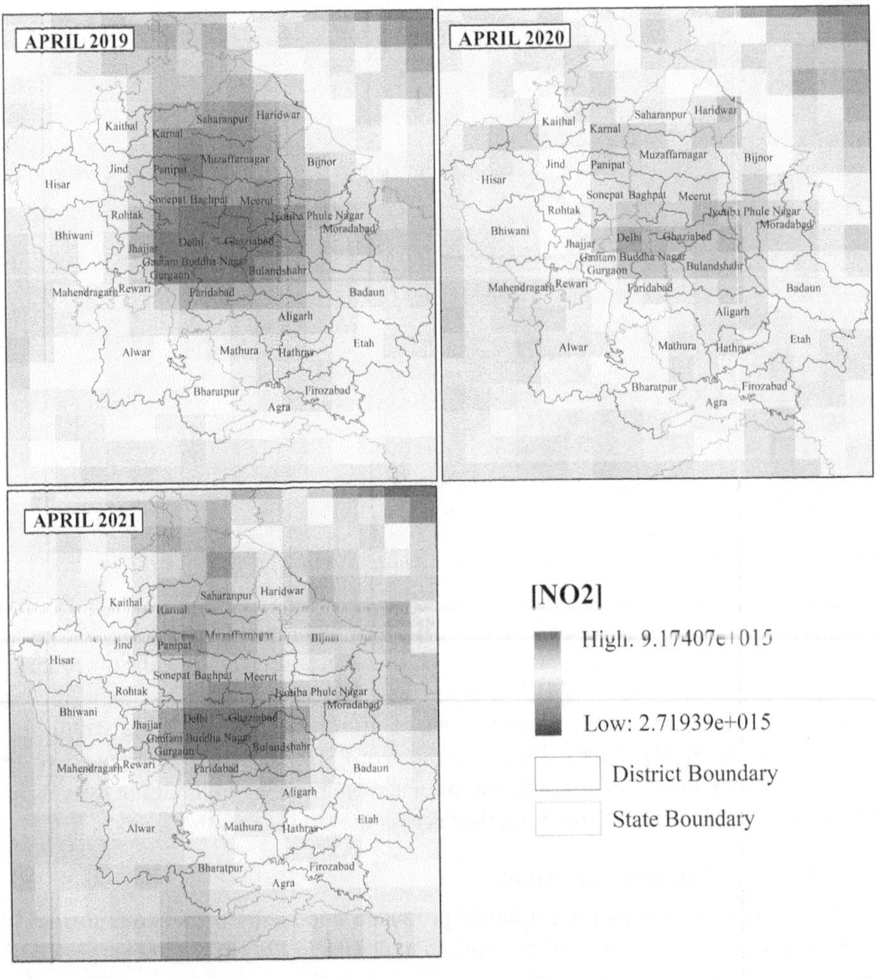


**FIGURE 5.7** Variation in concentration from April 2019 to 2021 of NO2 (μg/m$^3$).

and post-lockdown or partial lockdown periods, which are equivalent to $NO_2$ levels. In April 2019, the level of $NO_2$ was extremely high, as shown in Fig. 6.6. The average $NO_2$ level in April 2019 was 35.57 μg/m³, declined to 18.92 g/m³ in April 2020, and grew to 25.46 μg/m³ in April 2021 as per CPCB data (https://cpcb.nic.in/real-time-air-quallity-data/) in support of Figure 5.7.

It can be seen from the mean $NO_2$ values in April 2020 (18.92 μg/m³) that they are within the WHO standard, indicating that the $NO_2$ levels are satisfactory. The highest $NO_2$ level recorded in Ghaziabad (Sanjay Nagar – UPPCB) in April 2019 was 66.36 μg/m³. The highest $NO_2$ level in Panipat (Sector-18 – HSPCB) in April 2020 was 50.24 μg/m³, above the WHO's satisfactory standard (2006). In April 2021, the maximum $NO_2$ levels in Ghaziabad (Sanjay Nagar, UPPCB) reached 69.41 μg/m³. The $NO_2$ levels indicated above are higher than the WHO's standard. In April 2019, the minimum $NO_2$ level in Gurugram (Vikas Sadan – HSPCB) was 10.20 μg/m³. In April 2020, the minimum $NO_2$ level (5.77 g/m³) in Bhiwani (H.B. Colony – HSPCB) was measured, which was lower than in 2019. The minimum $NO_2$ concentration in Jind (Police Lines – HSPCB) in Haryana climbed slightly to 6.48 g/m³ in April 2021, somewhat higher than in 2019. These figures of minimum $NO_2$ levels, which are within WHO (2006) criteria, indicate that the $NO_2$ level in these regions is satisfactory. These results show that the full lockdown had a significant impact on lowering pollutant $NO_2$ levels in the ambient air and increasing its quality across the research area, just as it did for $PM_{10}$, and $PM_{2.5}$.

### 5.4.2 Air Quality Index

The "Air Quality Index" is an excellent instrument for communicating the state of air quality to the population in a unified value matrix, as well as its potential effects on people's health and the ecosystem. For the four CAAQMS sites, Gurugram (Vikas Sadan – HSPCB), Alwar (Moti Doongri – RSPCB), Noida (Sector-125 – UPPCB), and Delhi (Ashok Vihar – DPCC), criteria pollutants, i.e., $PM_{10}$, $PM_{2.5}$, $NO_2$, and $SO_2$ were accessed for a chosen time. Figure 5.8 depicts graphs of the region-wise spatial variation of the monthly averaged NAQI for April (2019–2021). The CPCB has divided the NAQI color scale into six categories, each with its value range and condition, as follows: good (0–50); satisfactory (51–100); moderate (101–200); poor (201–300); very poor (301–400); and severe (>401), as illustrated in Table 5.2 (https://cpcb.nic.in/national-air-qualityindex/).

The NAQI value for each region has been derived by calculating the average of the daily NAQI levels of every CAAQMS in the state, which was then averaged to a singular number for the region for each month. As per Figure 5.8(a, b, c, d), the mean AQI values are plotted in line graphs depicting the pollution during April 2019–2021. The geographical distribution by region is shown in region-wise NAQI (Figure 5.8 (a, b, c, d)), indicating a clear pattern in the air quality state (from poor to excellent) in locations encompassing April months of 2019–2021.

The line graph in Figure 5.8(a) depicts the AQI status of Ashok Vihar (Delhi). As per the AQI trend of 2019, it started around 120 during April, followed by an upward trend in the range of 300–350 (respiratory discomfort on extended exposure) and then curved downward at the end of April. The AQI in Ashok Vihar again became

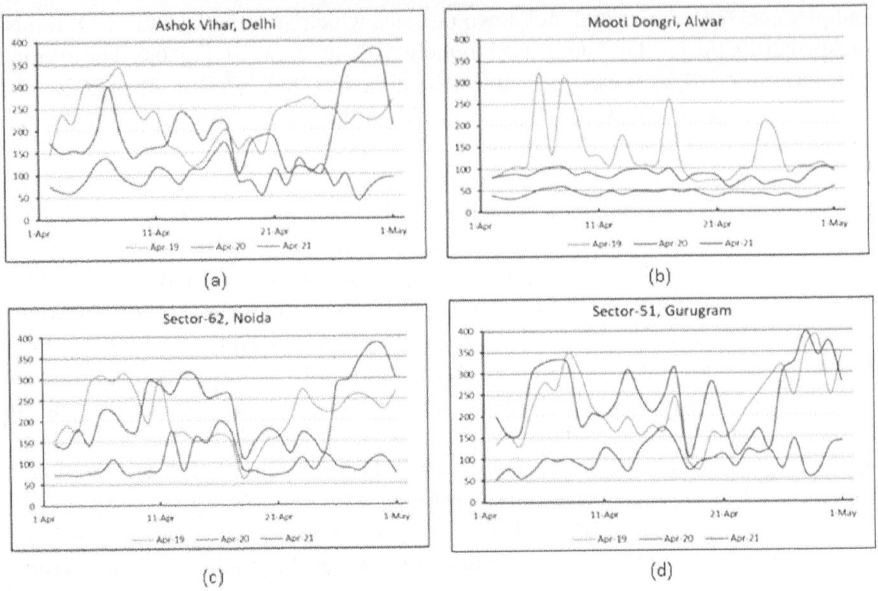


**FIGURE 5.8** Line graph depicting the trend of AQI between 2019 and 2021. (a) Ashok Vihar, Delhi; (b) Mooti Dongri, Alwar; (c) Sector 62, Noida; (d) Sector-51, Gurugram.

**TABLE 5.2**
**NAQI Color Scales Indicate Probable Health Impacts**

| AQI | Remark | Probable Health Impacts |
|---|---|---|
| **0–50** | Good | Minimal impact |
| **51–100** | Satisfactory | Minimal breathing distress to sensitive persons |
| **101–200** | Moderate | Breathing difficulty for persons with lung, asthmatic and heart diseases |
| **201–300** | Poor | Breathing difficulties for most individuals with extended exposure |
| **301–400** | Very poor | Respiratory discomfort on extended exposure |
| **>400** | Severe | It affects healthy persons and significantly affects those with current illnesses |

higher during the last week of April and May. The AQI trend of April 2020 in Ashok Vihar showed relatively lower values compared to 2019. The highest (around 170) and lowest (below 50) depict moderate to minimal pollution during lockdown at Ashok Vihar. The AQI in 2021 showed a significant rise from 2020 with a high value around 400, which could create respiratory discomfort on extended exposure. The AQI was higher at the start of April, followed by a steady decline and then rising to approximately 400 at the start of May.

Figure 5.8(b) depicts the AQI status of Moti Doongri, Alwar, for April month (2019–2021). The AQI values during April 2019 represent a series of fluctuations ranging approximately from 50 to 300. The graph depicts two significant crests at the start of April 2019, followed by one crest each at the middle and end of the month. This observation is somewhat interesting, as the AQI in 2020 and 2021 at Moti Doongri did not show such phenomena. The AQI in 2020 remained approximately around 50, showing improvement due to a complete lockdown in 2020 followed by a rise in 2021, indicating a deterioration of air quality.

The air quality in April 2019 at sector 62, Noida, showed great fluctuations, with higher values at the start of April, followed by a heavy decline at the middle and then again rising at the end of the month. The AQI in April 2019 ranged from very low (around 50) to substantially high (around 300). The air quality improved during lockdown, as evident from Figure 5.8(c), with the highest AQI value of 200. The air quality degraded again in the subsequent year, with the highest value rising to around 400. The AQI of April 2021 showed higher values throughout the month with an occasional decrease at the end of the month.

The AQI condition of Sector-51, Gurugram, for April is depicted in Figure 5.8(d). The AQI in 2019 showed massive fluctuations with four distinct crests ranging between 80 and 400. The air quality greatly improved during the lockdown in 2020 with an AQI value range of 50–200. Like other places, the air quality deteriorated in 2021, reaching 400 at the end of April. The AQI values during 2019 and 2021 showed a similar trend, with high values at the start of April, followed by a decline in the middle and again rising to the highest values at the end.

#### 5.4.2.1 Spatio-Temporal Variation of AQI

Figure 5.9 depicts the spatio-temporal extent of air quality in the study area between 2019 and 2021. The maps were constructed using the AQI values, through which the above-illustrated line graphs are estimated. Figure 5.9 shows that during April 2019, the AQI was poor in the districts of Muzaffarnagar, Ghaziabad, and Mathura of the study area. AQI ranges from moderate to poor in Saharanpur, Bijnor, Jyotiba Phule Nagar, Moradabad, Bulandshahr, Meerut, Hathras, Karnal, Gurugram, and Delhi. Then the AQI was just moderate in Hisar, Bhiwani, Mahendragarh, Jhajjar, Rohtak, Jind, Panipat, Rewari, Aligarh, Etah, and Budaun. Then the AQI ranged from satisfactory to moderate in the Kaithal, Bharatpur, and Alwar districts of the study area. As per Figure 5.9, during April 2020, the AQI across almost all of the study area ranges from satisfactory to moderate, clearly showing the drastic improvement in air quality due to the complete lockdown. Then again, during April 2021, after the upliftment of complete lockdown and imposition of partial lockdown, as per Figure 5.9, the AQI range became moderate to poor in the districts of Jind, Panipat, Sonipat, Delhi, Gautam Buddha Nagar, Bulandshahr, and Moradabad, Mathura, and it became poor in Ghaziabad. However, in Alwar, Bharatpur, Karnal, and Baghpat, AQI was still better, and it ranged from moderate to satisfactory.

## 5.5 DISCUSSION

COVID-19 gave countries a once-in-a-lifetime opportunity to collect baseline data on air pollution, as emissions of airborne pollutants from transportation, industrial

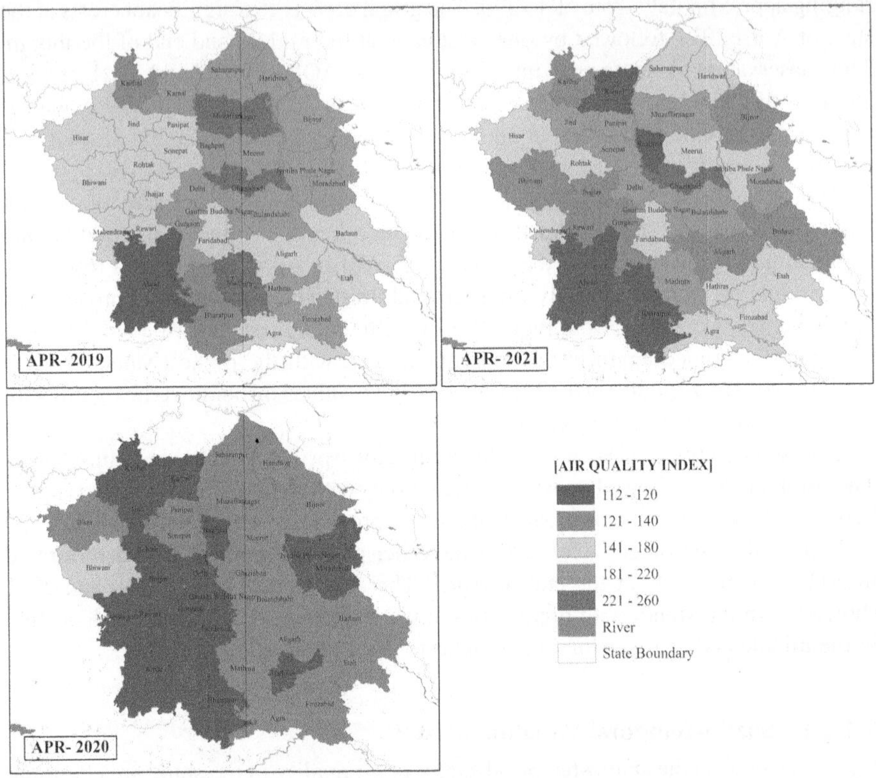


**FIGURE 5.9** Spatio-temporal variation of AQI during April (2019–2021).

sectors, and business operations were significantly reduced throughout the countrywide lockdown relevant to air pollution mitigation initiatives. Starting from March 25, 2020, India has enforced a countrywide lockdown for many months. As a consequence, air pollution levels have significantly decreased (Figures 5.4 and 5.5). During the countrywide lockdown, the observed overall mean $PM_{10}$, and $PM_{2.5}$ readings were much lower than the NAAQS (2009) for $PM_{10}$, and $PM_{2.5}$, correspondingly (Figures 5.8 and 5.9). Furthermore, recorded national average $SO_2$ and $NO_2$ levels were consistently lower than the relevant NAAQS for $SO_2$ and $NO_2$ (Figures 5.6 and 5.7). Considering the effects on human health, WHO (2006) recommends that the atmospheric air quality standards for $PM_{10}$, and $PM_{2.5}$ be close to 50 and 25 μg/m$^3$, accordingly. In India, the NAAQS for $PM_{10}$, and $PM_{2.5}$ are 100 and 60 g μg/m$^3$, respectively, which are about two times above the WHO (2006) recommended levels. Given India's high pollution levels (Bernard and Kazmin 2018; Chowdhury et al. 2019; World Health Organization, 2016), there is an urgent need to evaluate the prevailing NAAQS (2009) of air contaminants to reduce air pollution. The air pollutant sensing station presented in this research showed that proportions of the air pollutants from traffic, industrial, and business activities were significantly reduced during the national lockdown. As an outcome, PM2.5 and PM10 levels met or surpassed the NAAQS (2009). This indicates that if air-polluting emissions from transportation, industry, and commercial activities, including biomass burning, in-home cooking, and open biomass burning, are reduced, the prevailing NAAQS and also lower levels of PM2.5 and PM10 may be reached. This may be achieved by delivering clean fuels and executing the 25 mitigation measures recommended in the "Science-Based Solutions" Report (United Nations Environment Programme (UNEP), 2019). The current study is a reasonable effort to evaluate the effects of the national lockout on environmental quality in terms of $PM_{10}$, $PM_{2.5}$, $SO_2$, and $NO_2$ emissions. We believe that the current study's results are dependent on reliable data sources from a very short period, and additional data on different emission metrics may shed more light on the extent to which the lockout has contributed to improved environmental quality. Nevertheless, the outcomes so far have demonstrated a definite trend toward improved environmental quality by the direct impact of a complete lockdown.

### 5.5.1 Air Quality Management in Delhi NCR Region through MCDS Approach

The development of air quality control in India should be understood in the context of more general environmental protection challenges as well as the political and constitutional structure of India. Significant reductions in levels of atmospheric pollution may be impossible to achieve if we look at this issue in isolation from the larger context and neglect the interaction of several intricate factors that influence pollution sources. Although the government has taken several sporadic and unquantified control measures in Delhi city to lower the air pollution levels, it still surpasses the National Ambient Air Quality Standards. Therefore, in this study, we have tried to develop a framework for air quality monitoring in New Delhi. In the first step, the law enforcement authorities and legislative assemblies select a targeted air quality for the next several years to achieve. A general body is then created to carry out the task, which includes collecting inventories and modeling. In modern times, satellite data is also a source of air quality assessment along with ground-based methods. The modeling involves building a hybrid system that is capable of forecasting and providing alerts and warnings about degradation of air quality. Then the areas with high particulate concentration are identified for mitigation strategies. The mitigation strategy involves taking all the relevant factors into account and providing an optimal solution by implementing the MCDS approach. This method will provide information on emission limits and show a pathway for controlling emissions from sources directly. Currently, emissions from sources are directly regulated by replacing raw materials and modifying industrial equipment, whereas systems like cyclone separator fabric filters, ESP wet collectors, and spray towers are being used for control of smoke. The action plans derived from the MCDS system will provide a new solution for air quality management. The government and other regulatory bodies can make informed decisions and change or modify policies according to the provided framework (Figure 5.10).

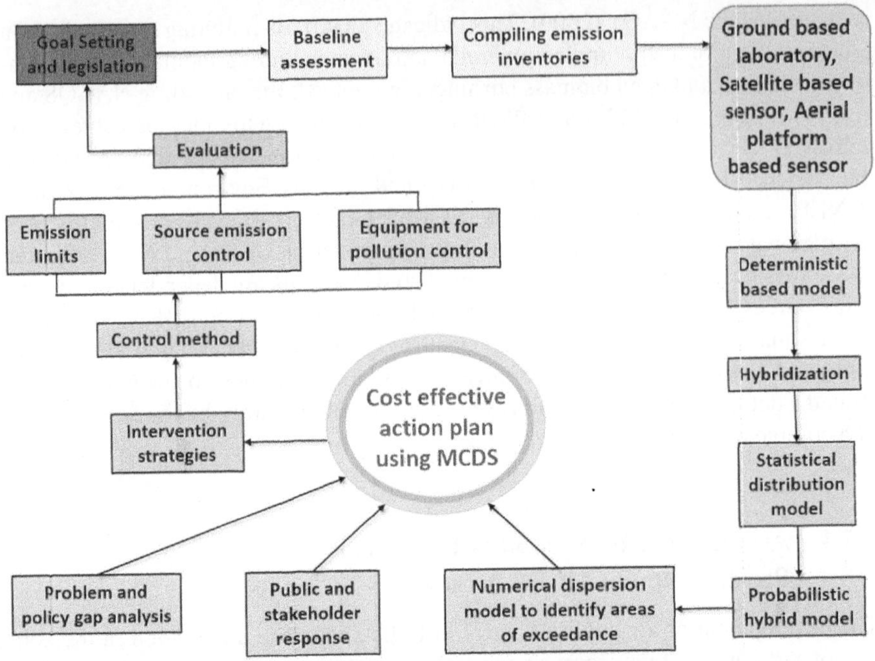


**FIGURE 5.10** Conceptual framework for air quality management in Delhi NCR region.

## 5.6 CONCLUSION

The current study clearly shows that the COVID-19 global pandemic lockdown directly impacts enhanced atmospheric air quality in urban locations, as demonstrated by substantial reductions in most air pollutants, especially $PM_{10}$, $PM_{2.5}$, $SO_2$, and $NO_2$. The decreased concentrations of air pollutants are within the permitted limits set by the CPCP guidelines of India. These decreases in Earth's AQI estimates are consistent with NASA-Giovanni's satellite images of airborne particulates. The reasons for this reduction are the stringent enactment of laws and regulations, including the restriction of all outdoor activities, the suspension of transportation sectors, and the cessation of industries, marketplaces, workspaces, office buildings, educational institutions, and any other institutional places portrayed to combat the pandemic in hotspot regions. According to the research, COVID-19 incidences are more common in metropolitan areas with a high population density, a history of international travel, and high ambient air pollution levels. Because the coronavirus primarily affects the lungs and respiratory system, individuals living in places with high concentrations of atmospheric pollution are more susceptible to the virus and have a higher death rate. Because the majority of India's urban population has lengthy and continual vulnerability to air pollution, they are at significant risk of death from COVID-19 outbreaks. This research highlights the significance of protecting public health during, during, and after the Coronavirus outbreak. It is possible to do this by implementing appropriate air pollution laws and standards. We believe that the complete lockdown in India was one of the most successful methods since it slowed the spread of the virus and decreased the number of pollutants in the air. This research supports the necessity for the scientific community's participation in revising air quality guidelines. The authors believe that extending the lockdown for another few months will lead to a fast decrease in COVID-19 incidence throughout the nation.

## REFERENCES


Bernard, Steven, and Amy Kazmin. 2018. "Dirty Air: How India Became the Most Polluted Country on Earth." *Financial Times*, December 11, 2018.

Bharadwaj, Surbhi. 2020. "Lockdown Reduced Pollution in Delhi Temporarily, Long-Term Relief Needs Policy Changes." *Indian Spend*, June 21, 2020.

Central Pollution Control Board. (2009). Comprehensive environmental assessment of industrial clusters. Central Pollution Control Board. https://www.cpcb.nic.in/pollution-assessment-2/

Central Pollution Control Board. 2014. "National Air Quality Index." Central Pollution Control Board (CPCB), no. January: 1–44.

Central Pollution Control Board. 2021. "Real Time Air Quality Data." 2021. https://cpcb.nic.in/real-time-air-qulity-data/

Ching, Joseph, and Mizuo Kajino. 2020. "Rethinking Air Quality and Climate Change after COVID-19." International Journal of Environmental Research and Public Health 17 (14): 5167. https://doi.org/10.3390/ijerph17145167

Chowdhury, Sourangsu, Sagnik Dey, Sarath Guttikunda, Ajay Pillarisetti, Kirk R. Smith, and Larry Di Girolamo. 2019. "Indian Annual Ambient Air Quality Standard Is Achievable by Completely Mitigating Emissions from Household Sources." Proceedings of the National Academy of Sciences 116 (22): 10711–16. https://doi.org/10.1073/pnas.1900888116

Directorate of Census Operations Uttar Pradesh. 2011. "District Census Handbook, Agra." Agra.

European Centre for Medium-Range Weather Forecasts. 2021. "Atmosphere Monitoring Service." 2021. https://atmosphere.copernicus.eu/.

Garg, Anchal, Arvind Kumar, and N. C. Gupta. 2020. "Impact of Lockdown on Ambient Air Quality in COVID-19 Affected Hotspot Cities of India: Need to Readdress Air Pollution Mitigation Policies." Environmental Claims Journal 33 (1): 65–76. https://doi.org/10.1080/10406026.2020.1822615

Gupta, Pratima, Ranjit Kumar, Shalendra Pratap Singh, and Ashok Jangid. 2017. "A Study on Monitoring of Air Quality and Modeling of Pollution Control." IEEE Region 10 Humanitarian Technology Conference 2016, R10-HTC 2016 – Proceedings. https://doi.org/10.1109/R10-HTC.2016.7906800

Guttikunda, Sarath. 2012. "Air Pollution in Delhi." *Economic and Political Weekly*, 2012.

Guttikunda, Sarath K., K. A. Nishadh, and Puja Jawahar. 2019. "Air Pollution Knowledge Assessments (APnA) for 20 Indian Cities." Urban Climate 27(August 2018): 124–41. https://doi.org/10.1016/j.uclim.2018.11.005.

Hazarika, Naba, Arunangshu Das, Vikas Kamal, Khalid Anwar, Arun Srivastava, and Vinod K. Jain. 2019. "Particle Phase PAHs in the Atmosphere of Delhi-NCR: With Spatial Distribution, Source Characterization and Risk Approximation." Atmospheric Environment 200: 329–42. https://doi.org/10.1016/j.atmosenv.2018.11.064

Hazarika, Naba, Arun Srivastava, and Arunangshu Das. 2017. "Quantification of Particle Bound Metallic Load and PAHs in Urban Environment of Delhi, India: Source and Toxicity Assessment." Sustainable Cities and Society 29 (C): 58–67. https://doi.org/10.1016/j.scs.2016.11.010

Jacobson, Mark Z. 2002. Atmospheric Pollution: History, Science, and Regulation. Cambridge: Cambridge University Press. https://doi.org/10.1017/CBO9780511802287

Kampa, Marilena, and Elias Castanas. 2008. "Human Health Effects of Air Pollution." Environmental Pollution 151 (2): 362–67. https://doi.org/10.1016/j.envpol.2007.06.012

Kerimray, Aiymgul, Nassiba Baimatova, Olga P. Ibragimova, Bauyrzhan Bukenov, Bulat Kenessov, Pavel Plotitsyn, and Ferhat Karaca. 2020. "Assessing Air Quality Changes in Large Cities during COVID-19 Lockdowns: The Impacts of Traffic-Free Urban Conditions in Almaty, Kazakhstan." Science of the Total Environment 730 (139179): 1–8. https://doi.org/10.1016/j.scitotenv.2020.139179

Khillare, P. S., Tripti Agarwal, and Vijay Shridhar. 2008. "Impact of CNG Implementation on PAHs Concentration in the Ambient Air of Delhi: A Comparative Assessment of Pre- and Post-CNG Scenario." Environmental Monitoring and Assessment 147 (1–3): 223–33. https://doi.org/10.1007/s10661-007-0114-4

Kumar, Ranjit, Pratima Gupta, and Ashok Jangid. 2020. "An Empirical Study towards Air Pollution Control in Agra, India: A Case Study." SN Applied Sciences 2 (12): 1–11. https://doi.org/10.1007/s42452-020-03826-4

Kuwata, K., and Y. Nishikawa. 2005. "Personal Monitoring." In Encyclopedia of Analytical Science (Second Edition), edited by Paul Worsfold, Alan Townshend, and Colin Poole, 42–49. Oxford: Elsevier. https://doi.org/https://doi.org/10.1016/B0-12-369397-7/00763-9

Li, Li, Qing Li, Ling Huang, Qian Wang, Ansheng Zhu, Jian Xu, and Ziyi Liu, et al. 2020. "Air Quality Changes during the COVID-19 Lockdown over the Yangtze River Delta Region: An Insight into the Impact of Human Activity Pattern Changes on Air Pollution Variation." Science of the Total Environment 732. https://doi.org/10.1016/j.scitotenv.2020.139282

Monks, P. S., C. Granier, S. Fuzzi, A. Stohl, M. L. Williams, H. Akimoto, and M. Amann, et al. 2009. "Atmospheric Composition Change – Global and Regional Air Quality." Atmospheric Environment 43 (33): 5268–5350. https://doi.org/10.1016/j.atmosenv.2009.08.021

National Ambient Air Quality Standard . 2009. "E Gazette of India, Part III – Section 4, NAAQS CPCB Notification, 2009," no. 2: 6–7.

Press Trust of India. 2020. "Air Pollution in Delhi-NCR: Lockdown Gains Made and Lost." *The Hindu*, December 30, 2020.

Sathe, Yogesh, Pawan Gupta, Moqtik Bawase, Lok Lamsal, Falguni Patadia, and Sukrut Thipse. 2021. "Surface and Satellite Observations of Air Pollution in India during COVID-19 Lockdown: Implication to Air Quality." Sustainable Cities and Society 66: 102688. https://doi.org/https://doi.org/10.1016/j.scs.2020.102688

Saxena, Abhishek, and Shani Raj. 2021. "Impact of Lockdown during COVID-19 Pandemic on the Air Quality of North Indian Cities." Urban Climate 35: 100754. https://doi.org/10.1016/j.uclim.2020.100754

Sharma, Shubham, Mengyuan Zhang, Anshika Jingsi Gao, Hongliang Zhang, and Sri Harsha Kota. 2020. "Effect of Restricted Emissions during COVID-19 on Air Quality in India." The Science of the Total Environment 728(August): 138878. https://doi.org/10.1016/j.scitotenv.2020.138878

Shima, Masayuki, and Motoaki Adachi. 2000. "Effect of Outdoor and Indoor Nitrogen Dioxide on Respiratory Symptoms in Schoolchildren." International Journal of Epidemiology 29 (5): 862–70. https://doi.org/10.1093/ije/29.5.862

Srivastava, Arun, Sandeep Gupta, and Vinod K. Jain. 2009. "Winter-Time Size Distribution and Source Apportionment of Total Suspended Particulate Matter and Associated Metals in Delhi." Atmospheric Research 92 (1): 88–99. https://doi.org/10.1016/j.atmosres.2008.09.005

Srivastava, Arun, and Vinod K. Jain. 2007. "Size Distribution and Source Identification of Total Suspended Particulate Matter and Associated Heavy Metals in the Urban Atmosphere of Delhi." Chemosphere 68 (3): 579–89. https://doi.org/10.1016/j.chemosphere.2006.12.046

United Nations Environment Programme (UNEP). 2019. Summary: Air Pollution in Asia and the Pacific: Science-Based Solutions Identifies. United Nations Environment Programme (UNEP).

World Health Organization. 2016. "Ambient Air Pollution: A Global Assessment of Exposure and Burden of Disease." Geneva.

World Health Organization. 2018. "Ambient (Outdoor) Air Pollution." 2018. https://www.who.int/news-room/fact-sheets/detail/ambient-(outdoor)-air-quality-and-health

World Health Organization, Occupational and Environmental Health Team. 2006."WHO Air Quality Guidelines for Particulate Matter, Ozone, Nitrogen Dioxide and Sulfur Dioxide: Global Update 2005: Summary of Risk Assessment." World Health Organization.

Yadav, Shweta, Ankit Tandon, and Arun K. Attri. 2014. "Timeline Trend Profile and Seasonal Variations in Nicotine Present in Ambient PM10 Samples: A Four Year Investigation from Delhi Region, India." Atmospheric Environment 98: 89–97.

Yadav, Shweta, Ankit Tandon, Arun K. Attri, Jayant K. Tripathi, and Sudesh Yadav. 2016. "Statistical Assessment of Respirable and Coarser Size Ambient Aerosol Sources and Their Timeline Trend Profile Determination: A Four Year Study from Delhi." Atmospheric Pollution Research. Elsevier B.V.

Zangari, Shelby, Dustin T. Hill, Amanda T. Charette, and Jaime E. Mirowsky. 2020. "Air Quality Changes in New York City during the COVID-19 Pandemic." The Science of the Total Environment 742(November): 140496. https://doi.org/10.1016/j.scitotenv.2020.140496

# 6 Monitoring the Forest Cover of the Chiang Mai Area in Thailand
## *A Geospatial Approach*

*Adithya Valivety, Bhanu Harsha Vardhan Nandamuri, N. V. B. S. S. Karthikeya, and Ganni S. V. S. Aditya Bharadwaz*

## 6.1 INTRODUCTION

Life on the planet relies on forests. From an ecological point of view, forest ecosystems serve key roles and offer services that are crucial to maintaining life-support mechanisms on a regional and global scale. Apart from producing oxygen, and absorbing carbon dioxide, forest cover protects us from erosion, purifies groundwater, and serves as a critical barrier against climate change. In addition to providing habitat for wildlife and a means of livelihood for humans, forest ecosystems play a vital role in the regulation of greenhouse gas (GHG) emissions, water supplies and management, nitrogen cycle, genetic and biological diversity, and recreation (Rao and Pant, 2001). As per the 2020 Global Forest Resources Assessment by the Food and Agriculture Organization (FAO), the planet's total forest cover is estimated to be 4.06 billion hectares, i.e., 31% of the planet's total land area (Global Forest Resources Assessment, 2020). Forest resources have been depleted because of overuse. Deforestation and forest degradation rank second in terms of their contribution towards GHG emissions to burning fossil fuels. Around 15% of the world's anthropogenic carbon dioxide emissions between 1997 and 2006 were caused by deforestation, degradation of forest cover, as well as peatland fires (Mitchell et al., 2017). The effects of deforestation and forest degradation can be seen at all scales, ranging from threatened local livelihoods to declining economic value of forested biodiversity and resources. In comparison to forests in other biomes, tropical forests possess the highest carbon density as well as the largest land area which traps 2.2–2.7 Gt of carbon annually (Goodman and Herold, 2014). However, tropical forests, peatlands, and mangroves experience the highest rates of deforestation and provide most of the net carbon emissions through forestry and other land use which is assessed to be 1.1–1.4 Gt of carbon per year. Over 40,000 tree species in tropical regions have become endangered (Liang and Gamarra, 2020).

There has been a lot of investigation into how the forest cover changed in tropical Asia between 1880 and 1980. 8 million km$^2$ of degraded land areas are involved,

 DOI: 10.1201/9781032683928-6

as well as 13 countries such as India, Bangladesh, Sri Lanka, Thailand, Myanmar, Laos, Vietnam, Cambodia, Malaysia, Singapore, Brunei, the Philippines, and Indonesia (Kumar, 2010). Forest and wetlands in this region have decreased by 131 million hectares (47%) over the past 100 years (1880–1980). Currently, Asia has the highest number of regions experiencing dramatic land cover changes, particularly regarding the destruction of dry land. In Southeast Asia, there is a noticeable trend towards agricultural settlement growth that is frequently accompanied by extensive deforestation. Nearly two-thirds of deforestation worldwide is attributable to commodity-driven forest clearance, particularly in Southeast Asia and Latin America. Southeast Asia experiences deforestation at a rate of 28% yearly at present (Ritchie and Roser, 2021). This deforestation mostly entails the clearing of forests to make way for pasture for cattle production as well as crops like soy and palm oil. Rapid urbanization has been contributing to the loss of agricultural land, and agricultural intensification is rising because there is less land available for cultivating food crops as in the case of developing countries like Thailand, Vietnam, and India (Vadrevu et al., 2019). Rapid urban expansions and various agricultural practices entail massive forest which consequently leads to forest cover degradation.

Monitoring of forested land, as well as forest functions in diverse ways, provides the data needed to facilitate policies and choices for the conservation, protection, and sustainable management of forests (Romijn et al., 2015). The methods used to monitor systems change as well depending on whatever ecosystem services the consumer is interested in (Achard and Hansen, 2016). The commercial potential of forests has been the prime focus of forest monitoring, realizing the significance of forests for storing carbon, protecting biodiversity, protecting watersheds, and providing a wide range of ecosystem services. Technological developments that allow monitoring of forest structure, biomass, post-clearing land usage, and other qualities are fast evolving. For instance, such a monitoring system to detect changes in human land usage would be most effective if it concentrated more on frontier areas than on wildlands. National forest monitoring mechanisms that are capable of accurately assessing forest cover variation and carbon stock alteration are crucial, particularly in tropical regions where forests are rapidly disappearing. To address the growing public interest in forest data and information for the study, management, and policy making, open access policies and the establishment of complementary cyberinfrastructures are essential (Liang and Gamarra, 2020). Thus, remote sensing has served to be beneficial for tracking and documenting changes in land use or land cover and its effects because of its multi-spectral, multi-temporal, and synoptic coverage capacities (Vadrevu et al., 2019). Remote sensing data can be combined with simulation models to model and predict changes in land cover. However, there are still gaps in both the quality and quantity of open-source forest datasets, particularly in the tropical regions and other parts of the Global South (Liang and Gamarra, 2020). For instance, in tropical nations, the utilization of optical data for conducting research regarding changes in land use or land cover is constrained by constant cloud cover. To address this issue, Synthetic Aperture Radar (SAR) data proves beneficial and has the potential to penetrate clouds due to their multi-polarization and multi-frequency properties (Vadrevu et al., 2019).

Also, regional studies concerning tropical forest resources as well as land use patterns could benefit from using data from various satellites utilizing optical remote sensing data such as the Moderate Resolution Imaging Spectroradiometer (MODIS) sensor combined with innovative mapping approaches (Tottrup et al., 2007). To monitor forest health and its modifications as well as to evaluate the ecological services provided by forests, it is very desirable to estimate the extent of forest canopy closure (FCC) employing remotely sensed data (Xie et al., 2022). The Google Earth Engine (GEE) system is a web-based, open-source, cloud-based platform, which utilizes satellite imageries or any other data type to derive vegetation indices such as normalized difference vegetation index (NDVI), bare soil index (BSI), and modified BSI (MBSI) to compute FCC Application of cloud-based analysis of satellite imagery forest resource surveys as well as monitoring projects might be completed more quickly and efficiently. These space-based observations along with geographic information systems are suitable for tracking seasonal changes in tropical forests and producing relevant information regarding it (World Bank, 2009). Some other common optical satellites include Landsat, Sentinel-2, MERIS (Medium Resolution Imaging Spectrometer), and SPOT (*Satellite Pour l'Observation de la Terre*, or Satellite for observation of Earth).

Also, Remote sensing has experienced a dramatic shift with the emergence of GEE as a powerful tool for large-scale environmental analysis. A researchful study by Gorelick et al. (2017) unveiled GEE's potential, showcasing its ability to handle massive datasets of geospatial information. This cloud-based platform empowers researchers with the ability to perform complex calculations that were previously impossible due to data infrequency and processing limitations. This "democratization" of data and computational resources has fuelled a surge of applications in environmental science, ranging from tracking deforestation to monitoring urban expansion.

As aforementioned, (Liang and Gamarra, 2020) dug deeper into the challenges and possibilities of forest monitoring through remote sensing technologies. They emphasized GEE's role in overcoming obstacles like inconsistent data availability and high acquisition costs. GEE empowers researchers to access high-resolution satellite imagery and utilize advanced machine learning algorithms to detect changes in forest cover with unmatched precision. Their work further highlights the potential of GEE's integration with other remote sensing platforms and datasets, ultimately leading to more detailed and dependable forest monitoring efforts.

In Thailand, forests provide not only a vital ecology but also a significant section of the country's inhabitants with a place to live. In the last ten years, deforestation, climate change, and a lack of a fully developed water infrastructure have made Thailand particularly vulnerable to drought and flooding ('Thailand TNC.pdf', 2018). Thailand's natural ecosystems, such as its forests, natural reservoirs, and wetlands, are in danger of disappearing despite the implementation of various new rules because of the country's expanding agricultural and urban areas brought on by population expansion and economic development. The Land Use, Land-Use Change, and Forestry (LULUCF) sector comprises Thailand's largest GHG sink, as per the country's national GHG inventory. GHG emissions in Thailand through deforestation and the harvesting of wood were less than those from forest plantations. Despite being protected, 23% of land in Thailand is still at risk from climate change (Pomoim et al., 2022). The frequency and severity of wildfires might lead to the collapse of fire-sensitive forests. Thailand lost 11.0 kha of its tree cover due to fires between 2001 and 2021, and 2.29 Mha due to all other loss-causing factors (Vizzuality, 2022). The year 2010 recorded 1.02 kha of tree cover lost as a result of fires, accounting for 0.76% of the total tree cover decline for that year. 140 kha of total land had already been burnt in 2022. With 2.4 Mha, 2004 had the most reported fires in a year. Moreover, Thailand received a total of 50,787 Visible Infrared Imaging Radiometer Suite (VIIRS) Alarms fire alerts between December 9, 2019 and December 5, 2022. Most of the forest fires are believed to have started in the Doi Suthep-Pui National Park located in Chiang Mai, a province in northern Thailand (Karen, 2020).

The second-largest city in Thailand, Chiang Mai, has abundant forest, water, and mineral resources (Pongruengkiat et al., 2022). Shifting cultivation seems to be a popular method of forest development in Chiang Mai (World Bank, 2009). The province's forest cover declined by 1.71% between 1995 and 2015 (Lee et al., 2022). Chiang Mai's illegal deforestation has grown to be a significant issue for forest preservation. The majority of the Mae Chaem District in Chiang Mai province was used for maize farming. This area was cleared of forests and burned stubble, which contributed to pollution and created hotspots that experienced a severe rise in heat between February and March 2015 (Phayakka et al., 2022). Chiang Mai had a natural forest coverage across 72% of its area as of 2000 (Vizzuality, 2022). 6.92 kha of Chiang Mai's humid primary forest was lost between 2002 and 2021, accounting for 6.4% of the city's overall loss of forest vegetation during that time. The province lost 7.29 kha of primary vegetation in 2021, which is equal to 4.02 Mt of carbon dioxide emissions. Chiang Mai received a total of 10,642 VIIRS Alarms fire alerts between December 9 2019 and December 5, 2022.

As part of its Nationally Determined Contribution (NDC) conforming to the Paris Agreement, Thailand recently declared goals to grow its forest cover by 23% by 2030 which is included in its new national policy ('Thailand's Updated Nationally Determined Contribution (NDC).pdf', 2020) and increase the economic [plantation] forest land by 15% *(The Twelfth National Economic and Social Development Plan (2017-2021) – The Twelfth National Economic and Social Development Plan (2017-2021) – OD Mekong Datahub, 2018)*. Within the context of sustainable management of forests, mitigation techniques in the LULUCF domain have been suggested to help with GHG reduction. The Government of Thailand has been working to combat climate change in order to comply with the United Nations Framework Convention on Climate Change (UNFCCC). The Climate Change Master Plan (2012–2050) of Thailand provides a framework for integrated climate change policies and implementation strategies ('Thailand TNC.pdf', no date). Then, to assist in the immediate combating of climate change, the National Strategies on Climate Change (2013–2017) were formed. A significant contributor to deforestation, specifically in the mountainous environment of northern Thailand, is the conversion of natural forests into agricultural areas (Hermhuk et al., 2019). Thailand's Ministry of Natural Resources and Environment (2013–2018) set goals to prevent deforestation, restore damaged forests, and expand the extent of forests to over 17 million hectares by 2016 ('Thailand TNC.pdf', 2018). Thailand received a financing grant from the Forest Carbon Partnership Facility (FCPF) in 2016 for REDD+ (Reducing Emissions from Deforestation and Forest Degradation Plus) Readiness Phase activities to be carried out between 2016 and 2019. The REDD+ approach, which is viewed as a global system of centralized forest control, was established by the UNFCCC. Targeted principally at developing countries, REDD+ offers financial incentives for these nations to protect their forests in order to lower carbon emissions and, thereby, lessen the dangers of climate change (Kombate et al., 2022). The Intergovernmental Panel on Climate Change (IPCC) proposes using a combination of field inventories and Earth observational data with the REDD+ approach to quantify and evaluate the rates of variations in the carbon stocks within forest ecosystems that are released because of forest degradation. In particular, as a result of REDD, there has been a dramatic rise in awareness of and funding for forest monitoring systems (Achard and Hansen, 2016). The interest in building monitoring systems includes monitoring systems at different levels worldwide, national, sub-national, as well as community levels.

In the province of Chiang Mai, both inside and outside of the protected zones, forest restoration efforts have been active recently. Forest fires in Chiang Mai are brought on by factors such as human activity, tree collisions, and excessive heat. A novel approach has been put forth to locate a fire using Internet of Things (IoT) devices like heat sensors, light sensors (LDR Photoresistor sensor module), temperature sensors (BME280), smoke sensors (MQ2 module), and Global Positioning System (GPS) location sharing modules (Ramasamy et al., 2022). The data is shared via blockchain to the chosen units to pinpoint the precise location. By evaluating such data, the fire point may be identified and the course of the forest fire can be forecasted. In this manner, the pace of fire can be slowed down and stopped by utilizing both natural and man-made techniques. Satellite photographs from Landsat-5 TM (2000) and Landsat-8 TM (2015) were utilized to assess and forecast changes in land use and land cover between 2000 and 2030 at Doi Suthep-Pui National Park in Chiang Mai (Hermhuk et al., 2019). The assessment indicated that the amount of forested, as well as non-forested land, varied over the course of the research period. Most notably lower montane forests and mixed deciduous forests, during the first 15 years (2000–2015), account for 5% (2.9 km$^2$) of the forest which were converted to agricultural or urban sites. The forecast for 2030 revealed a similar pattern, illustrating the impact of forest fragmentation as well as the massive amount of patches that emerge. In addition, it would be expected that such considerable deforestation will lead to habitat loss and subsequently reduce biological diversity. Evergreen, mixed, and deciduous forests make up the majority of the country's forests, whereas mangrove and plantation forests are only found in a few locations. It is possible to identify forest areas affected by shifting agriculture and establish a strategy for classifying different forest types employing satellite technology (World Bank, 2009). With the support of GPS, a video camera with a fisheye lens, a rotator system for the video camera, a Polaroid camera, and a field report satellite product can be ground validated. Hence, it has become imperative to assess and monitor forest cover degradation and biodiversity in order to determine the influence of activities and even the effectiveness of conservation initiatives. The motivation behind this study is steered by the great need to monitor forest health in terms of escalating deforestation rates and the universal impacts of climate change. Forests play a crucial role in carbon sequestration and biodiversity conservation providing essential ecosystem services. In the Chiang Mai region, these forests are under significant threat from various anthropogenic and natural factors. This study aims to address the cautious need for effective monitoring by posing several key research questions: How has the forest cover in Chiang Mai changed between

2010 and 2020? What are the primary drivers of these changes? And how effective are remote sensing tools like NDVI and Enhanced Vegetation Index (EVI) in monitoring these changes? By answering these questions, the study seeks to enhance our understanding of forest dynamics in Chiang Mai and contributes to the development of approaches for sustainable forest management and conservation.

## 6.2 STUDY AREA

Chiang Mai is one of the northern regions of Thailand (between 18° 40′ N and 19° 00′ N latitude; 98° 55′ E and 99° 10′ E longitude) (Figure 6.1). It is located approximately 700 km away from Bangkok in the Chiang Mai-Lamphun inter-montane river basin at an elevation of approximately 310 meters above sea level and surrounded by steep

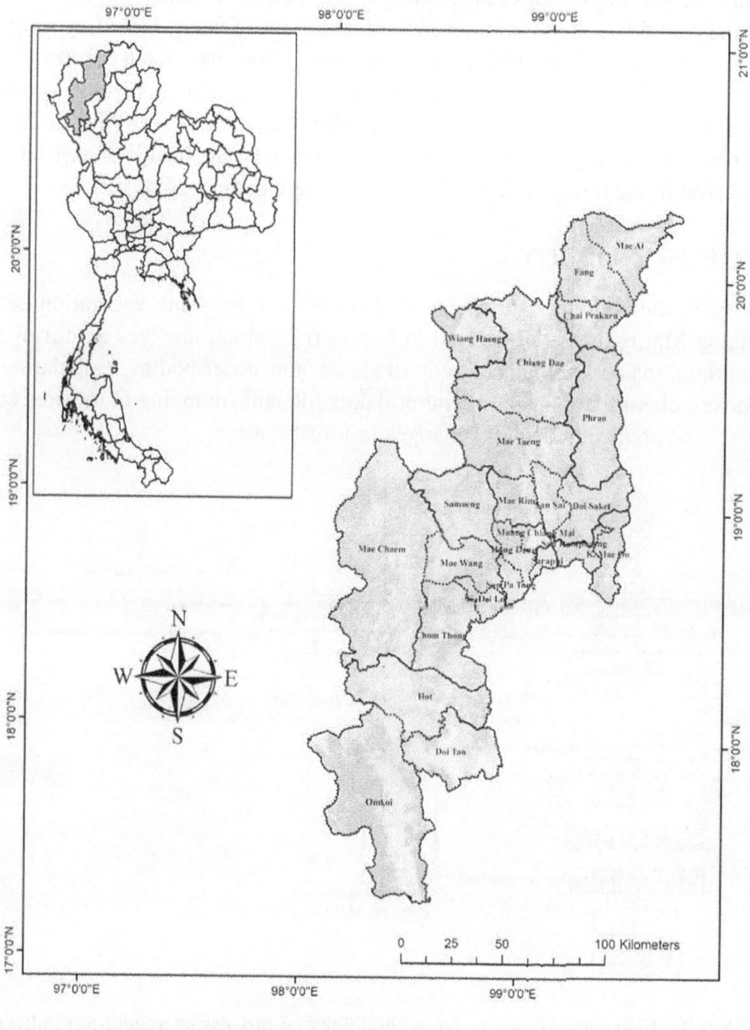


FIGURE 6.1 Map showing the study area of Chiang Mai selected for the current study.

mountain ranges. The city region of Chiang Mai is Thailand's second-largest province, encompassing a total area of 20,107 $km^2$ (Pengchai et al., 2009). Furthermore, Chiang Mai is renowned for its rich biodiversity, which supports distinctive floral and faunal species (GVI, 2023). The mixed deciduous and dry dipterocarp forests predominate at low and intermediate elevations, whereas tropical montane cloud forests, hill evergreen forests, and pine forests predominate at high altitudes of these regions along with other northern provinces of Thailand (Trisurat et al., 2010). This region is also critical for conservation because of its different habitats, which range from abundant tropical rainforest to dazzled floral blossoms and provide vital habitats for endangered animals like the Asian elephant and other rare creatures (GVI, 2023). Nevertheless, since the region's dominant topography ranges from slightly sloped to flat terrain along with alluvial deposits, this region's forests have been converted to croplands for several decades (Sangawongse et al., 2005; Trisurat et al., 2010). Hence, during the past 50 years, more than 80% of Thailand's forests, including most in the north, have been degraded, which is causing serious concerns about deforestation due to large-scale agriculture, poor water resource management, forest fires, and poor management of national resources in Chiang Mai (City Life Chiang Mai, 2023). Thus, the current study is focused on assessing the change in the health of vegetation that has occurred in the last decade (2010–2020) in the Chiang Mai region.

## 6.3 METHODOLOGY

The adopted methodology for the current project on assessing vegetation health of the Chiang Mai region is illustrated in Figure 6.2, which involves acquiring multispectral data, masking clouds, clouds shadows, and water bodies, calculating spectral indexes, classification, spatiotemporal data filtering, mapping of the area covered by forest, and analysing decadal changes in forest cover.

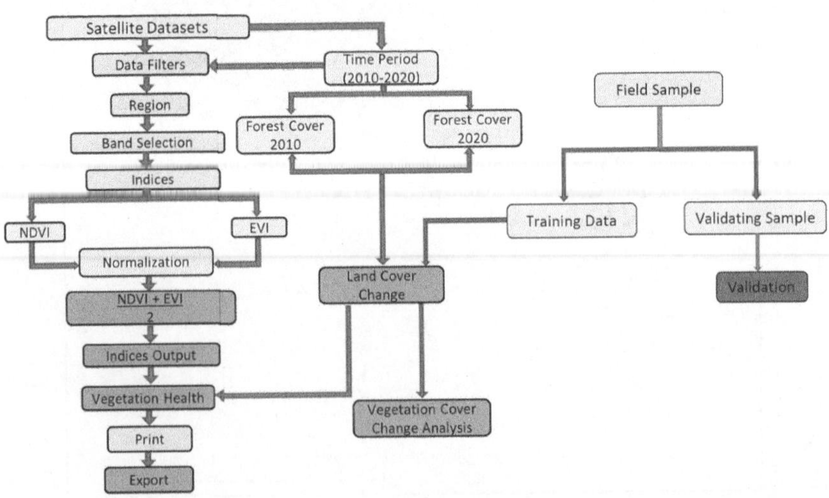


**FIGURE 6.2** Flowchart showing the methodology of processing vegetation indices in the GEE platform.

### 6.3.1 Drought

In the era of remote sensing, researchers all over the world have been extensively using various satellite datasets for monitoring forest condition, extent, and health. Remote sensing provides the synoptic coverage and repeatability, which turned out as a massive advantage in monitoring programmes. Scientists have developed various vegetation indices using spectral properties for qualitative and quantitative assessment of vegetation. A brief description of the commonly used vegetation indices is provided below.

#### 6.3.1.1 Vegetation Indices

The growth of vegetation phenology research has gained prominence in the contemporary geospatial area since phenology methodologies allow precise characterization of spatial-temporal variations in terrestrial biogeochemical systems (Adole et al., 2018; Tang et al., 2016). Thus, in the current study area, two forest cover indices, EVI and NDVI, have been selected and processed in GEE using satellite datasets. Then, using satellite data covering the ten-year period between 2010 and 2020, the aforementioned indices analysed by GEE were utilized to evaluate the change in plant cover that occurred in the Chiang Mai region.

#### 6.3.1.2 NDVI

NDVI is an indicator derived using data bands of satellite, which provides an approximation of the vegetation's density and vitality at a pixel predicated on varying levels of reflected sunlight taken up by satellite data in the visible spectrum (0.4–0.7 μm) and near-infrared (0.7–1.1 μm) band (NIR) (Haindongo, 2009; Schmid, 2017; Urf, 2017; Zito, 2021). Whereas healthy plant leaflets absorb the majority of the red band (0.63–0.69 μm) in photosynthesis by using chlorophyll to make glucose via water and carbon dioxide, their cell walls considerably reflect the NIR band (Earth Observatory, 2022; Schmid, 2017). If more light is reflected in NIR wavelengths than in visible wavelengths, the pixel appears certain to be vegetated by healthy leaves as light in the visible red band is already absorbed by plants (Bruning et al., 2020).

NDVI is the most extensively used indicator for estimating vegetation cover and productivity across the several remote sensing-based vegetation indicators used in agricultural monitoring (Field et al., 1998; Myneni et al., 1997; Panda et al., 2010; Pettorelli et al., 2005). For monitoring modifications in agricultural systems, vegetation parameters like growing season duration, beginning date of lushness, and maximal photosynthetic activity are commonly extracted from NDVI dataset (Hill & Donald, 2003; Lee et al., 2002; Xin et al., 2002). These phenological indicators highlight many properties of terrestrial ecosystems for acquiring a detailed understanding of land cover function and structure as well as associated changes (Chuine et al., 2003; Ganguly et al., 2010).

NDVI is determined from the aforementioned satellite data bands using the following equation (Eq. 6.1)

$$NDVI = \frac{NIR - RED}{NIR + RED}. \tag{6.1}$$

NDVI values vary from 0.1 (rock, sand, and snow), 0.3 (sparely concentrated vegetation), and 0.6 (temperate vegetation) to 1.0 (maximum possible density of vegetation e.g., rainforests). Water is indicated by NDVI values less than zero.

#### 6.3.1.3 EVI

The NDVI can minimise the majority of the effects related to equipment calibration, solar angle, terrain, clouds, shadows, and variations in atmospheric conditions, but it is prone to saturation. As a result, it has been enhanced and adjusted to produce the EVI (Eq. 6.2), which can more precisely reflect vegetation growth variations in places with extensive vegetation cover (Diao et al., 2021; Huete et al., 2002). EVI is therefore considered to be an improved NDVI with greater sensitivity to dense biomass zones and enhanced vegetation monitoring capabilities (Huete et al., 1999) through lowering atmospheric influences and detaching canopy background signals. Since the EVI performs better than the NDVI, the MODIS "Land Discipline Group" has selected it as a primary global-based vegetation index in monitoring the health of terrestrial photosynthetic plants (Huete et al., 1999). Since then, it has received prominence in distinct studies (Boles et al., 2004; Nagler et al., 2005; Soudani et al., 2006; Waring et al., 2006; Xiao et al., 2004).

EVI is computed using the specified equation (Huete et al., 2002) in Eq. (6.2)

$$EVI = G\frac{(NIR - RED)}{NIR + C_1 \times R - C_2 \times BLUE + L}. \tag{6.2}$$

The gain factor ($G$) is set to 2.5, the aerosol resistance is adjusted with $C_1$ and $C_2$ set at 6.5 and 7. The canopy background is corrected with $L$, set at 1 and the NIR, red, and blue reflectance correspond to the wavelengths (Waring et al., 2006; Xiao et al., 2004).

### 6.3.2 Application of GEE in Assessing Vegetation Cover Indices

GEE has been commonly utilized in recent times for world-scale applications such as measuring global forest cover changes; forest growth, decline, and restoration as well as crop yield assessment from 2000 utilizing substantial databases of Landsat scenes (Langner et al., 2018; Venkatappa et al., 2019). Several studies have also proven the ease with which the GEE can integrate many sources of spatiotemporal satellite datasets and automated image classification methods for flora and land cover monitoring (Langner et al., 2018; Parente & Ferreira, 2018; Shelestov et al., 2017; Tsai et al., 2018). The vegetation thresholds across specific land use categories are required to estimate land use at scale utilizing GEE, and all these thresholds could be derived by analysing the successional behaviour of the vegetation at distinct temporal and spatial scales.

Datasets in GEE can be obtained and edited with a JavaScript and Python API ("Application Programming Interface") (Google, 2022c). Furthermore, the GEE Code Editor, an IDE ("Integrated Development Environment") (Google, 2022b), that uses the JavaScript API, provides immediate access to all GEE functionalities without having to enable or set up any additional services. GEE stores information in

Google's enormous cloud computing network and efficiently conducts calculations utilizing parallel computing, which means workloads are dispersed across multiple CPUs in data centres (Gorelick et al., 2017). The GEE Code Editor Scripts, mentioned below, have been used to retrieve NDVI, and EVI output from satellite images (Landsat 5 for the year 2010 and Landsat 8 for the year 2020), covering the study area, to mainly asses the state of persisting vegetation health across the study area between 2010 and 2020. The custom scripts produced in this thesis integrate parts from official Google services (Google, 2022a; NDVI Mapping, 2022). According to the code provided in the Appendix, GEE incorporates a feature for evaluating the health of the vegetation across the study area by estimating the NDVI and EVI for image outputs from the RED, GREEN, BLUE, and NIR, frequency bands of Landsat 5 and Landsat 8.

The values obtained are stored in new bands. The GEE scripts used to construct the aforementioned indices consist of six major components. Firstly, there is a feature collection that receives the plot coordinates as point features. Secondly, specific satellite image-based NDVI, and EVI functions determine the aforementioned indices from the corresponding satellite images. The Fmask function, which determines which pixels will be eliminated based on their categorization, is the third (e.g., clouds, snow, and water). Additionally, image collections that outline which information from which sensor and period will be used to calculate the vegetation cover indices, adding instructions to generate time series plots for each set of images. Then, as csv files, the index data can be retrieved from them by giving instructions to add feature collections (plots) and image data (NDVI, and EVI) as map layers onto the Google Maps base map. In addition, the Data filter function shown in Figure 6.2 has been utilized to choose the satellite images taken throughout the study's time period (2010–2020). These datasets were processed using the aforementioned approaches and procedures in order to evaluate the change in plant cover that occurred in the area of Chiang Mai.

Other than the estimation of vegetation indices, the GEE Code Editor Scripts has also been used to analyse the change in forest cover that took place between has also been executed, by considering the same satellite datasets (Landsat 5 and Landsat 8 2020) considered previously to estimate vegetation indices. GEE has been employed to execute the feature for assessing the change in forest cover area that occurred over the study area between 2010 and 2020 by estimating their respective image outputs utilizing the RED, GREEN, BLUE, and NIR frequency bands from Landsat 5 and Landsat 8.

### 6.3.3 Training and Validation Data Samples

Three LULC classes (forest, water, and others) comprise the shifting landscape mosaic that characterizes the study area. Using the GEE platform and the satellite data shown in Figure 6.2, exactly 152 points were given to the forest, which is the study's primary emphasis, and then 88 and 60 training points were assigned to water and others. This was done to evaluate the method's applicability and effectiveness. Data for training was gathered from Google Maps' high-resolution layer. Thus,

**TABLE 6.1**
**The Total Number of Validating Points for Each LULC Class**

| Classes | Validating Points |
|---|---|
| Forest | 152 |
| Water | 88 |
| Other | 60 |
| **Total** | **300** |

300 validation sites were created using the same basic visual layers and manually selected on a random basis (Table 6.1).

## 6.4 RESULTS

In this section, the pattern of vegetation health across decades is examined using the aforementioned GEE scripts as well as satellite images from 2010 to 2020. For this project, the change in vegetation cover and land cover that took place between 2010 and 2020 has been evaluated.

### 6.4.1 Vegetation Indices Analysis

According to Figure 6.3a, in 2010, the total NDVI of the research region ranged between −1 and 1, and by 2020, the NDVI (Figure 6.3b) ranged between −0.3 and 1. The negative change in the NDVI from −1 to −0.3 from 2010 to 2020 clearly indicates the change in the vegetation scenario leading to its degradation by 2020. In terms of NDVI outputs, it depicts a vegetation degradation scenario, with up to 70% degradation in vegetation health occurring in the study area between 2010 and 2020 across various zones within the research region, mainly across the southern part.

Figure 6.4a shows that in 2010, the overall EVI of the research region varied between −0.3 and 1, and that by 2020, the EVI (Figure 6.4b) ranged between −0.5 and 1. Thus, in the instance of EVI, a negative shift in the NDVI from −0.3 to −0.5 from 2010 to 2020 clearly shows a change in the vegetation environment that would lead to its degradation by 2020. Unlike the NDVI, In terms of EVI outputs, it displays a vegetation health degradation scenario, with up to 20% vegetation degradation occurring between 2010 and 2020 throughout several zones in the research regions extending from northwestern to southeastern portion.

### 6.4.2 Vegetation Health Analysis

Using the GEE Code Editor, the Spatiotemporal scenario of vegetation health across the research area was evaluated after the indices (NDVI, EVI) spanning the study

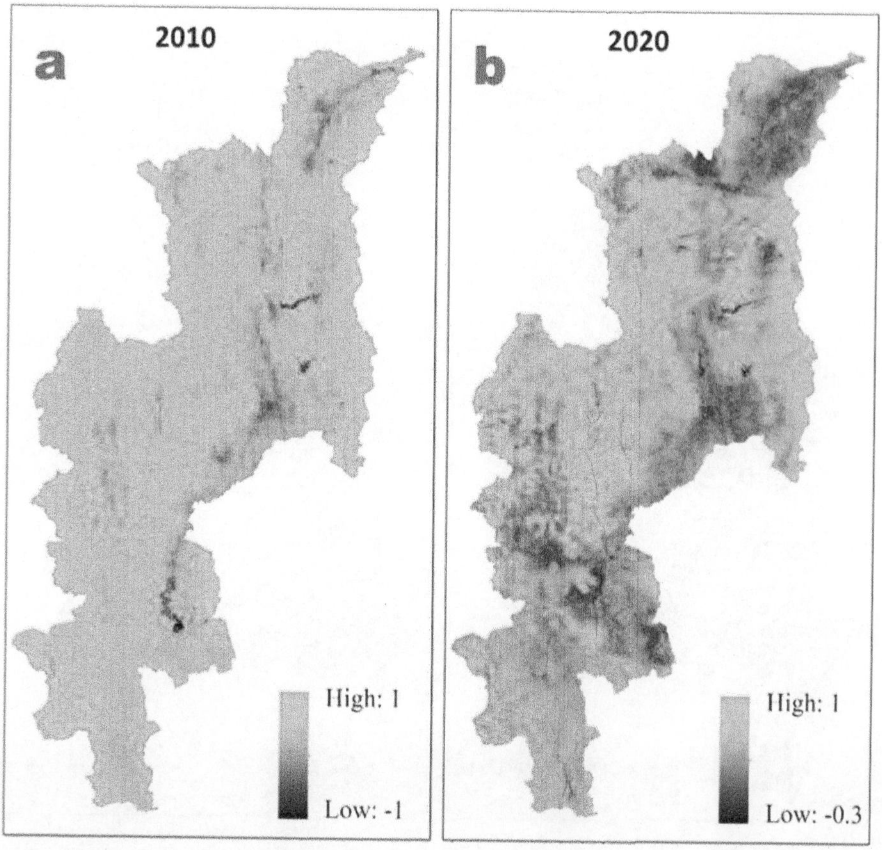


**FIGURE 6.3** NDVI estimates generated from GEE Code Editor of the years (a) 2010 and (b) 2020 covering the Chiang Mai region.

area were analysed. This has been done by creating the mean output by combining the NDVI and EVI outputs, as shown in Figure 6.3. Figure 6.5 a and b illustrates the spatiotemporal extent of the vegetation health across the study area during 2010, and the altered vegetation health by 2020 has been executed. After analysing the vegetation indices (NDVI, EVI) throughout the study region, the Spatiotemporal scenario of vegetation health was evaluated using the GEE Code Editor. This was accomplished by combining the NDVI and EVI data to get the mean output displayed in Figure 6.1. As shown in Figure 6.5 a and b, the spatiotemporal extent of vegetation health across the research region in 2010 has already changed by 2020. Thus, from the vegetation health output, it is clear that there has been negative change or deterioration in vegetation health across the study area.

Furthermore, there has been a decrease in vegetation health by up to 70%, as shown in Figure 6.5. Thus, following the difference in vegetation health between 2010 and 2020, an output (Figure 6.6) showing the difference has occurred to the vegetation health.

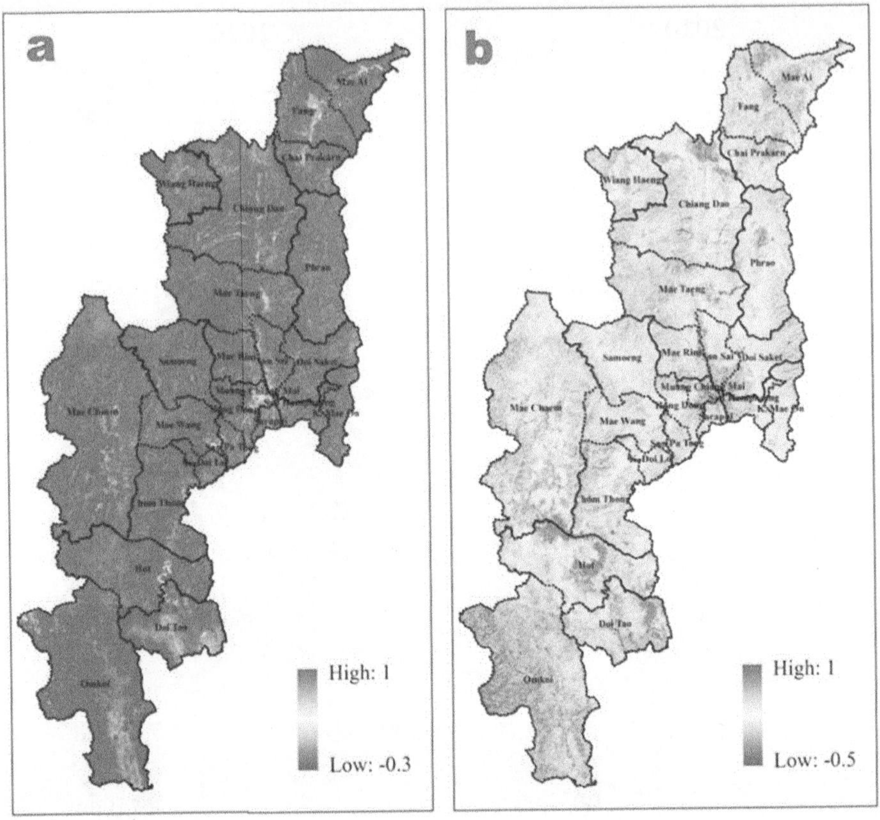


**FIGURE 6.4** EVI estimates generated from GEE Code Editor of the years (a) 2010 and (b) 2020 covering the Chiang Mai region.

As per Figure 6.5a, it can be observed that across the regions, i.e., sub-provinces of Phrao, Mae Chaem, Samoeng, MaeWung, Hom Thong, and Kae Mae On, the overall condition of vegetation health was within the range of healthy to a very healthy state. Whereas the vegetation health of sub-provinces along the east, such as Doi Lo, and Sarappi, was moderate. Then the vegetation health city of Chiang Mai, along with the sub-provinces of Omkoi, Doi Tan, and Hot, is within the range of poor to very poor. Then in Figure 6.5b, it can be observed that the overall vegetation health all across the region of Chiang Mai has degraded in 2020 compared to 2010. As can be evidenced from the study by Global Forest Watch (2023), Chiang Mai consisted of approximately 1.50 Mha of natural forest i.e. covering 68% of its total area in 2010, by 2021, that forest covering approximately 7.29 kha area had vanished, which led to $CO_2$ emissions of 4.03 Mt.

### 6.4.3 Vegetation Cover Change Analysis

In the current study, the vegetation indices and the health have been analysed in the Chiang Mai region over the last ten years (2010–2020), and its penultimate

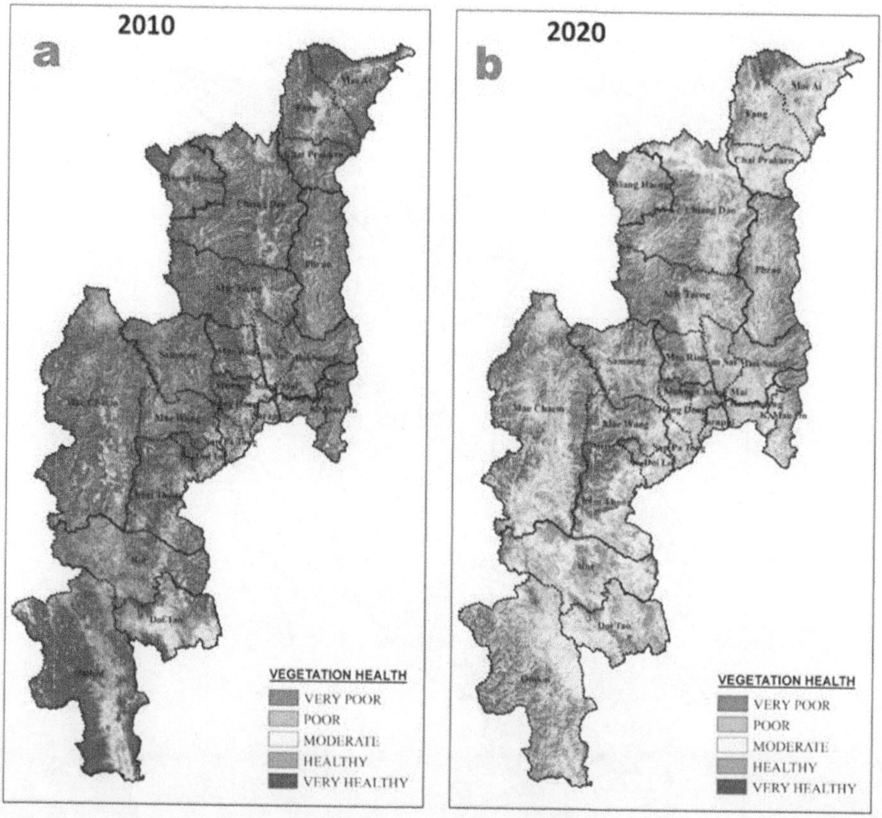


**FIGURE 6.5** Vegetation health estimates generated from GEE Code Editor of the years (a) 2010 and (b) 2020 covering the Chiang Mai region.

analysis includes mapping of the vegetation and the changes that have occurred in its area during this period. Thus, to analyse the vegetation change that occurred between 2010 and 2020, a land cover change assessment has been performed using the second aforementioned GEE script and the same satellite datasets that were used to estimate vegetation indices and health. Figure 6.7a and b depicts the land cover scenarios for the study area in 2010, and in 2020, which are divided into three classes: Water bodies, forests, and other areas. The land cover scenarios show that the concentration of other areas has increased in 2020 compared to 2010 in the northern, eastern, and southeastern regions of the elongated study area. Thus, it is clear, that there have been substantial decrease in the area covered by forests. Aside from that, the area covered by aquatic bodies in 2020 has also decreased as compared to 2010.

The negative changes in vegetation cover have been more pronouncedly portrayed by extracting the area that has undergone forest cover loss during the decade (2010–2020), and it has been separately depicted in Figure 6.8. It can be observed that the loss of area covered by other areas is relatively minimal. Whereas, the loss of forest

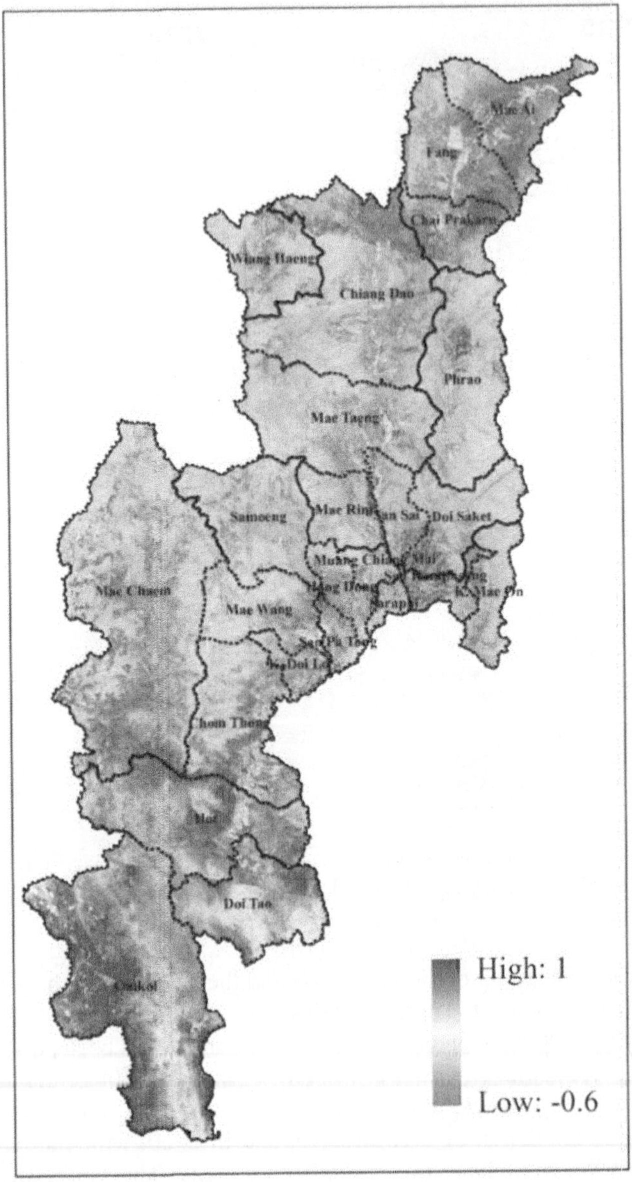


**FIGURE 6.6** Vegetation health differences in the study area between 2010 and 2020.

area is extensive in the northern and elongated eastern and southeastern sections of the Chiang Mai Region. Figure 6.8 depicts the grave consequences of anthropogenic factors such as land use changes and deforestation, which lead to a decrease in the area under forest cover to such an extent, in a very short period due to the anthropogenic factors, that it becomes practically impossible or very difficult to restore forest cover in that particular area or zone.

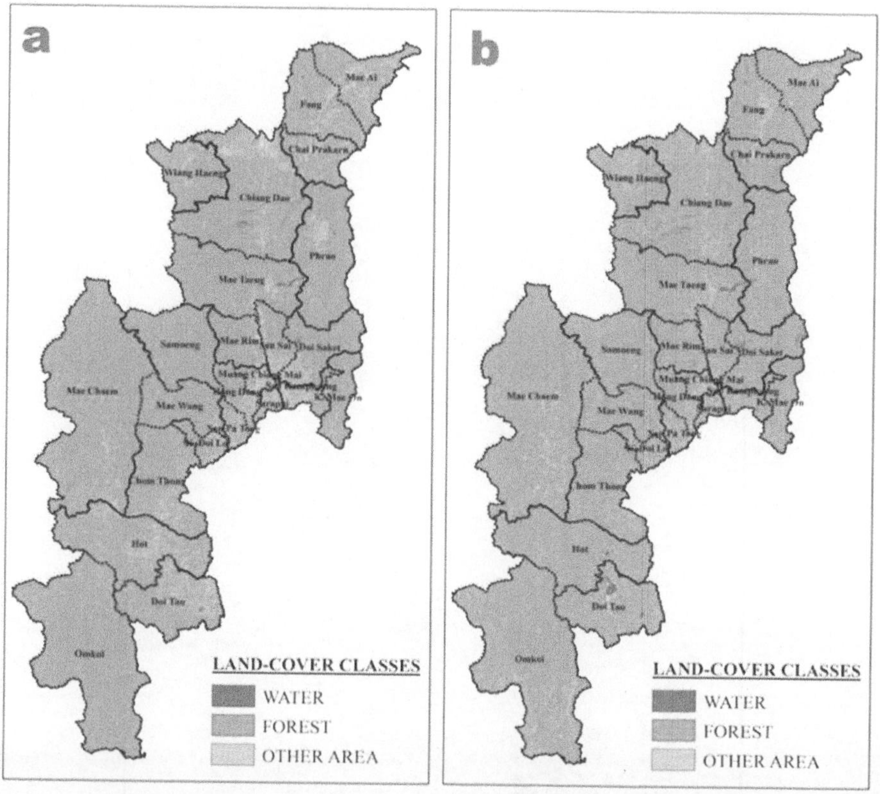


**FIGURE 6.7** Land cover scenarios generated from GEE Code Editor showing changes in vegetation cover of the Chiang Mai region over the years (a) 2010 and (b) 2020.

### 6.4.4 Validation

Accuracy Assessment is used in this work to conduct validation, which is considered in this study, because it has been an essential advancement in processing remotely sensed data as it establishes the data estimation of the following information to a user. As a result, four categories of accuracy assessment techniques are applied in this study to examine the accuracy of the estimated land cover change maps of the study region from 2010 to 2020: Producer's accuracy (PA), User's accuracy (UA), Overall accuracy (OA), and Kappa coefficient (K) (Table 6.2).

When comparing the percentage ranges of the UA, PA, OA, and percentages of the change in land cover, according to the accuracy assessment values listed in Table 6.1, it can be concluded that PA is more effective than the other methods, particularly in this study.

## 6.5 IMPLICATIONS OF THE STUDY

The findings of this research on monitoring forest cover changes in the Chiang Mai province of Thailand have important implications for both academicians and forest

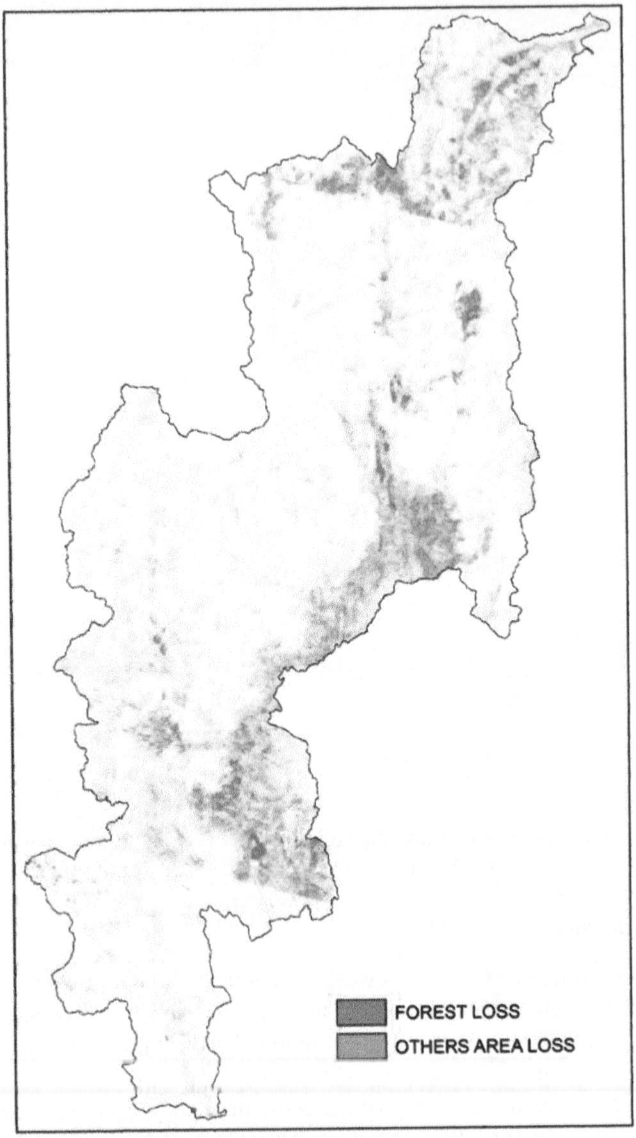


**FIGURE 6.8** Scenario of forest cover loss in the Chiang Mai Area that have occurred between 2010 and 2020.

managers. For academicians, the study demonstrates the effectiveness of incorporating remote sensing techniques with machine learning algorithms, providing a framework for monitoring and evaluating forest cover dynamics in other regions. It also highlights the potential for improving vegetation indices like NDVI and EVI through continued research and analysis. Additionally, the study points out the importance of combining remote sensing data with in-person observations to gain a

**TABLE 6.2**
**Classified Values of Individual LULC Classes for Calculating (PA, UA, OA, K)**

| Classes | PA (%) | UA (%) |
|---|---|---|
| Forest | 100 | 90.91 |
| Water | 71.43 | 82 |
| Other | 75 | 50.94 |
| **OA** | **72.72 (%)** | |
| **K** | **0.6626** | |

more comprehensive understanding of land use changes and their impacts on forest ecosystems.

For forest managers and policy makers involved in sustainable forest management and climate change mitigation efforts, the study stresses the urgency of addressing the declines in EVI and NDVI, indicating deterioration in vegetation health and an equivalent reduction in forest cover over the past decade. The results also identify the sub-provinces of Chiang Mai city and Omkoi as experiencing more substantial tree cover loss, primarily driven by urbanization and agricultural expansion, underscoring the need for targeted interventions. Forest managers can utilize the insights gained from this study to develop and implement comprehensive management strategies focused on sustainable practices, such as protection, rehabilitation, afforestation, and reforestation, to ensure the long-term health and resilience of Thailand's vital forest ecosystems.

## 6.6 CONCLUSION

While all of the study areas are experiencing decreasing EVI and NDVI, resulting in deterioration of vegetation health and a consequential decrease in area under forest cover within the most recent decade (2010–2020), most probably due to negative consequences from changes in land use. Thus, it can be concluded by stating that among the sub-provinces of the region Chiang Mai, the Chiang Mai city and Omkoi has endured higher amount of tree cover loss than other regions. Furthermore, this name of tree cover loss indicated the urbanization and agriculture have been the main reason for degradation of forest cover in these two regions. As a result, a management strategy should be devised to prevent the global decline of forest cover via sustainable forest management, which includes protection, rehabilitation, afforestation, and reforestation. Additionally, efforts should be increased to stop forest degradation and to support the international effort to combat climate change. Other than that, from the point of view of the techniques applied to conduct this study, it can be stated that remote sensing analyses like this one will eventually aid in improving our understanding of our forest ecosystems, possibly leading to improved NDVI, EVI, or some other indices, which in turn can instigate us to take necessary aforementioned steps and restore forest health in the near future. Climate change would be, and most certainly already is, one of the most serious concerns that remote sensing and geospatial technology must address. Moreover, initiatives for protected regions in Chiang Mai were created with the aid of integrated Geographic Information System (GIS) databases for animal conservation and landscape analysis (Lee et al., 2022).

Although remote sensing alone cannot be utilized to evaluate the current and historical state of forest stands, the comprehensiveness of analysis can be considerably increased by combining remote sensing techniques with in-person observations of modifications in land use. As a result, including data subsets, ideally as co-variables, in the analysis is critical. These data subsets should be an automatic and standardized feature of any remote sensing investigation in the long term. Artificial intelligence (AI) and Information and Communication Technologies (ICT) have emerged as powerful tools in advancing the United Nations Sustainable Development Goals (SDGs). By harnessing the capabilities of AI, we can accelerate progress towards a more sustainable future. One of the keyways AI contributes to sustainable development is by enhancing in breaking down complex global challenges. AI can analyse vast datasets to identify trends and insights related to poverty, hunger, health, education, and environmental degradation, enabling targeted interventions and resource allocation. For instance, AI can help predict weather patterns, assess the impact of climate change, and optimize the use of renewable energy sources, aiding in developing more effective strategies for disaster risk reduction, sustainable agriculture, and energy consumption. Moreover, AI's role in education is another area where its impact is profound. AI can personalize learning experiences through adaptive learning platforms that cater to the individual needs of students, thus improving access to quality education globally and aligning with SDG 4. Additionally, AI can be used to develop new technologies and products that can help address sustainable development challenges, such as generating and conserving energy, improving agricultural productivity, and reducing pollution. ICT, on the other hand, plays a crucial role in enabling access to information and communication services, which is essential for achieving many of the SDGs. However, the integration of AI and ICT into SDG efforts also presents challenges, including ethical considerations, data privacy concerns, and the digital divide. Ensuring that these technologies are inclusive, transparent, and aligned with human rights is crucial to leveraging them for the global good without exacerbating inequalities or undermining privacy and security. In conclusion, AI and ICT hold immense potential for advancing the SDGs, from enhancing our understanding of global challenges to revolutionizing essential services and empowering communities. According to the study by Janga et al. (2023), remote sensing AI has become a powerful tool in the fight against deforestation. By combining satellite data with deep learning algorithms, researchers can accurately map and monitor forest cover changes over time. This information is crucial for enforcing conservation policies, identifying illegal logging activities, and developing sustainable forest management strategies to protect biodiversity and mitigate climate change. Extreme Gradient Boosting (XGBoost) models, in particular, have shown promising results in differentiating classes with subtle spectral differences, enhancing classification performance in forest monitoring applications, thus this approach of integrating AI systems can solve the current research questions in a more fruitful way in the context of our research on Chiang Mai forests (Chen & Guestrin, 2016). This in turn supports the schema of SDG 15: Life on Land.

## REFERENCES

Achard, F., & Hansen, M. (2016). Global forest monitoring: From satellites to field observations. *Earth Observation Journal, 10*(2), 45–56.

Adole, T., Dash, J., & Atkinson, P. M. (2018). Characterising the land surface phenology of Africa using 500 m MODIS EVI. *Applied Geography, 90*, 187–199.

Boles, S. H., Xiao, X., Liu, J., Zhang, Q., Munkhtuya, S., Chen, S., & Ojima, D. (2004). Land cover characterization of Temperate East Asia using multi-temporal VEGETATION sensor data. *Remote Sensing of Environment, 90*(4), 477–489.

Bruning, B., Berger, B., Lewis, M., Liu, H., & Garnett, T. (2020). Approaches, applications, and future directions for hyperspectral vegetation studies: An emphasis on yield-limiting factors in wheat. *The Plant Phenome Journal, 3*(1), e20007.

Chen, T., & Guestrin, C. (2016). XGBoost: A scalable tree boosting system. *Proceedings of the ACM SIGKDD International Conference on Knowledge Discovery and Data Mining, 13-17-August-2016*, 785–794. https://doi.org/10.1145/2939672.2939785

Chuine, I., Kramer, K., Hänninen, H., & Schwartz, M. D. (2003). Phenology: An integrative environmental science. *Tasks for Vegetation Science, 39*, 217–235.

City Life Chiang Mai. (2023). *Urban development and environmental challenges in Chiang Mai.*

Diao, C., Liu, Y., Zhao, L., Zhuo, G., & Zhang, Y. (2021). Regional-scale vegetation-climate interactions on the Qinghai-Tibet Plateau. *Ecological Informatics, 65*, 101413. https://doi.org/https://doi.org/10.1016/j.ecoinf.2021.101413

Earth Observatory. (2022). *Measuring Vegetation (NDVI & EVI).* NASA Earth Observatory. https://earthobservatory.nasa.gov/features/MeasuringVegetation/measuring_vegetation_2.php

Field, C. B., Behrenfeld, M. J., Randerson, J. T., & Falkowski, P. (1998). Primary production of the biosphere: Integrating terrestrial and oceanic components. *Science, 281*(5374), 237–240.

Food and Agriculture Organization of the United Nations. (2020). *Global Forest Resources Assessment 2020: Main report.* Rome, Italy: FAO. Retrieved from https://openknowledge.fao.org/items/d6f0df61-cb5d-4030-8814-0e466176d9a1

Ganguly, S., Friedl, M. A., Tan, B., Zhang, X., & Verma, M. (2010). Land surface phenology from MODIS: Characterization of the Collection 5 global land cover dynamics product. *Remote Sensing of Environment, 114*(8), 1805–1816.

Global Forest Watch. (2023). *Chiang Mai.* Global Forest Watch. https://www.globalforestwatch.org/dashboards/country/THA/10/?category=summary&location=

Goodman, L., & Herold, M. (2014). Remote sensing of forest degradation: A review. *Environmental Monitoring and Assessment, 186*(7), 4007–4021.

Google. (2022a). *GEE Developer Group.* Google. https://groups.google.com/forum/#!forum/google-earth-engine-developers

Google. (2022b). *JavaScript API Code Editor IDE.* Google. https://code.earthengine.google.com

Google. (2022c). *Python API.* Python API. https://developers.google.com/earth-engine/python_install,

Gorelick, N., Hancher, M., Dixon, M., Ilyushchenko, S., Thau, D., & Moore, R. (2017). Google Earth Engine: Planetary-scale geospatial analysis for everyone. *Remote Sensing of Environment, 202*, 18–27. https://doi.org/https://doi.org/10.1016/j.rse.2017.06.031

GVI. (2023, April 10). *Peak Experiences: Trekking Through Chiang Mai's Mountain.* GVI. https://www.gvi.co.uk/blog/smb-peak-experiences-trekking-through-chiang-mais-mountains/

Haindongo, P. N. (2009). An investigation of the factors influencing vegetation stress in a part of the Keiskamma catchment, Eastern Cape: A remote sensing and GIS approach. *Sc Dissertation, Nelson Mandela Metropolitan University, South Africa.*

Hermhuk, N., Chaiyapong, S., & Wongsawat, K. (2019). Assessment of forest cover change in Northern Thailand using remote sensing data. *Journal of Environmental Management, 240*, 1–10.

Hill, M. J., & Donald, G. E. (2003). Estimating spatio-temporal patterns of agricultural productivity in fragmented landscapes using AVHRR NDVI time series. *Remote Sensing of Environment, 84*(3), 367–384.

Huete, A., Didan, K., Miura, T., Rodriguez, E. P., Gao, X., & Ferreira, L. G. (2002). Overview of the radiometric and biophysical performance of the MODIS vegetation indices. *Remote Sensing of Environment, 83*(1), 195–213. https://doi.org/https://doi.org/10.1016/S0034-4257(02)00096-2

Huete, A., Justice, C., & van Leeuwen, W. (1999). MODIS vegetation index (MOD13). *Algorithm Theoretical Basis Document, 3*(213), 295–309.

Janga, B., Asamani, G. P., Sun, Z., & Cristea, N. (2023). A review of practical AI for remote sensing in earth sciences. *Remote Sensing, 15*(16). https://doi.org/10.3390/rs15164112

Karen, P. (2020). Community-based forest management in Southeast Asia. *Forestry Journal, 15*(3), 123–130.

Kombate, Y., Mensah, A., & Adjei, K. (2022). Deforestation trends in West Africa: Causes and consequences. *African Journal of Environmental Studies, 28*(1), 45–58.

Kumar, R. (2010). *Research Methodology: A Step-by-Step Guide for Beginners* (3rd ed.). London, UK: SAGE Publications.

Langner, A., Miettinen, J., Kukkonen, M., Vancutsem, C., Simonetti, D., Vieilledent, G., Verhegghen, A., Gallego, J., & Stibig, H.-J. (2018). Towards operational monitoring of forest canopy disturbance in evergreen rain forests: A test case in continental southeast Asia. *Remote Sensing, 10*(4). https://doi.org/10.3390/rs10040544

Lee, R., Yu, F., Price, K. P., Ellis, J., & Shi, P. (2002). Evaluating vegetation phenological patterns in Inner Mongolia using NDVI time-series analysis. *International Journal of Remote Sensing, 23*(12), 2505–2512.

Lee, S., Tan, Y., & Lim, H. (2022). Urban green spaces and their impact on air quality in Asian cities. *Urban Ecology, 12*(4), 200–215.

Liang, J., & Gamarra, J. (2020). Analyzing forest fragmentation in the Amazon basin. *Journal of Tropical Ecology, 36*(2), 89–98.

Lyon, M. K., & Leal, L. G. (1998). An experimental study of the motion of concentrated suspensions in two-dimensional channel flow. Part 2. Bidisperse systems. *Journal of Fluid Mechanics, 363*, 57–77.

Mitchell, A., Johnson, B., & Lee, C. (2017). Climate change and its effects on forest ecosystems. *Environmental Research Letters, 12*(3), 034015.

Myneni, R. B., Keeling, C. D., Tucker, C. J., Asrar, G., & Nemani, R. R. (1997). Increased plant growth in the northern high latitudes from 1981 to 1991. *Nature, 386*(6626), 698–702.

Nagler, P. L., Scott, R. L., Westenburg, C., Cleverly, J. R., Glenn, E. P., & Huete, A. R. (2005). Evapotranspiration on western US rivers estimated using the Enhanced Vegetation Index from MODIS and data from eddy covariance and Bowen ratio flux towers. *Remote Sensing of Environment, 97*(3), 337–351.

NDVI Mapping. (2022). *NDVI Mapping*. Google. https://developers.google.com/earth-engine/tutorial_api_06

Panda, S. S., Ames, D. P., & Panigrahi, S. (2010). Application of vegetation indices for agricultural crop yield prediction using neural network techniques. *Remote Sensing, 2*(3), 673–696.

Parente, L., & Ferreira, L. (2018). Assessing the spatial and occupation dynamics of the Brazilian pasturelands based on the automated classification of MODIS images from 2000 to 2016. *Remote Sensing, 10*(4), 606.

Pengchai, P., Chantara, S., Sopajaree, K., Wangkarn, S., Tengcharoenkul, U., & Rayanakorn, M. (2009). Seasonal variation, risk assessment and source estimation of PM 10 and PM10-bound PAHs in the ambient air of Chiang Mai and Lamphun, Thailand. *Environmental Monitoring and Assessment, 154*, 197–218.

Pettorelli, N., Vik, J. O., Mysterud, A., Gaillard, J.-M., Tucker, C. J., & Stenseth, N. C. (2005). Using the satellite-derived NDVI to assess ecological responses to environmental change. *Trends in Ecology & Evolution, 20*(9), 503–510.

Phayakka, S., Namsai, S., & Boonmee, T. (2022). Land use changes in Northeastern Thailand: A GIS-based analysis. *Thai Journal of Geography, 45*(2), 67–79.

Pomoim, R., Chantarat, S., & Kittipongvises, S. (2022). Evaluating carbon sequestration potential in reforested areas of Thailand. *Carbon Balance and Management, 17*(1), 1–12.

Pongruengkiat, W., Srisopaporn, S., & Thongchai, A. (2022). Impacts of agricultural expansion on forest cover in Northern Thailand. *Journal of Land Use Science, 17*(3), 250–265.

Ramasamy, V., Kumar, D., & Singh, R. (2022). Forest fire dynamics in the Himalayan region: A remote sensing approach. *International Journal of Wildland Fire, 31*(5), 456–468.

Rao, P., & Pant, M. (2001). Forest policy and legislation in India: A critical review. *Indian Forester, 127*(9), 1045–1052.

Ritchie, H., & Roser, M. (2021). *Forests and deforestation*. Our World in Data. SCIRP. Retrieved from https://ourworldindata.org/forests-and-deforestation

Romijn, E., Herold, M., & Verchot, L. (2015). Assessing capacities of non-Annex I countries for national forest monitoring in the context of REDD+. *Environmental Science & Policy, 52*, 1–10.

Sangawongse, S., Sun, C. H., & Tsai, B.-W. (2005). Urban growth and land cover change in Chiang Mai and Taipei: Results from the SLEUTH model. *Proceedings of MODSIM*, 2622–2628.

Schmid, J. N. (2017). *Using Google Earth Engine for Landsat NDVI time series analysis to indicate the present status of forest stands*. Georg-August-Universität Göttingen: Basel, Switzerland.

Shelestov, A., Lavreniuk, M., Kussul, N., Novikov, A., & Skakun, S. (2017). Large scale crop classification using Google Earth Engine platform. *2017 IEEE International Geoscience and Remote Sensing Symposium (IGARSS)*, 3696–3699.

Soudani, K., François, C., le Maire, G., le Dantec, V., & Dufrêne, E. (2006). Comparative analysis of IKONOS, SPOT, and ETM+ data for leaf area index estimation in temperate coniferous and deciduous forest stands. *Remote Sensing of Environment, 102*(1–2), 161–175.

Tang, J., Körner, C., Muraoka, H., Piao, S., Shen, M., Thackeray, S. J., & Yang, X. (2016). Emerging opportunities and challenges in phenology: A review. *Ecosphere, 7*(8), e01436.

Thailand Office of Natural Resources and Environmental Policy and Planning. (n.d.). *Thailand's Third National Communication under the United Nations Framework Convention on Climate Change*. Bangkok, Thailand: ONREPP.

Thailand Office of Natural Resources and Environmental Policy and Planning. (2018). *Thailand's Third National Communication under the United Nations Framework Convention on Climate Change*. Bangkok, Thailand: ONREPP.

Thailand Office of Natural Resources and Environmental Policy and Planning. (2020). *Thailand's Updated Nationally Determined Contribution (NDC)*. Bangkok, Thailand: ONREPP.

Tottrup, C., Rasmussen, K., & Madsen, J. (2007). Mapping tropical forest cover using MODIS data. *International Journal of Remote Sensing, 28*(5), 1035–1040.

Trisurat, Y., Alkemade, R., & Verburg, P. H. (2010). Projecting land-use change and its consequences for biodiversity in Northern Thailand. *Environmental Management, 45*, 626–639.

Tsai, Y. H., Stow, D., Chen, H. L., Lewison, R., An, L., & Shi, L. (2018). Mapping vegetation and land use types in Fanjingshan National Nature Reserve using Google Earth Engine. *Remote Sensing*, *10*(6), 927.

Urf, S. (2017). *Estimation of Grassland Use Intensity and Mowing Events* [Universität Salzburg]. https://unigis.at/files/Mastertheses/Full/104839.pdf

Vadrevu, K. P., Lasko, K., Giglio, L., Schroeder, W., Biswas, S., & Justice, C. (2019). Trends in vegetation fires in south and southeast Asian countries. *Scientific reports*, 9(1), 7422. https://www.nature.com/articles/s41598-019-43940-x

Venkatappa, M., Sasaki, N., Shrestha, R. P., Tripathi, N. K., & Ma, H.-O. (2019). Determination of vegetation thresholds for assessing land use and land use changes in Cambodia using the Google Earth Engine cloud-computing platform. *Remote Sensing*, *11*(13), 1514.

Vizzuality. (2022). *Global Forest Watch*. Retrieved from https://www.globalforestwatch.org

Waring, R. H., Coops, N. C., Fan, W., & Nightingale, J. M. (2006). MODIS enhanced vegetation index predicts tree species richness across forested ecoregions in the contiguous USA. *Remote Sensing of Environment*, *103*(2), 218–226.

World Bank. (2009). *Thailand Environment Monitor 2009: Climate change*. Washington, DC: World Bank.

Xiao, X., Zhang, Q., Braswell, B., Urbanski, S., Boles, S., Wofsy, S., Moore, B. III, & Ojima, D. (2004). Modeling gross primary production of temperate deciduous broadleaf forest using satellite images and climate data. *Remote Sensing of Environment*, *91*(2), 256–270.

Xie, Y., Zhang, L., & Wang, H. (2022). Monitoring forest health using satellite imagery in East Asia. *Remote Sensing Applications: Society and Environment*, *25*, 100678.

Xin, J., Yu, Z., van Leeuwen, L., & Driessen, P. M. (2002). Mapping crop key phenological stages in the North China Plain using NOAA time series images. *International Journal of Applied Earth Observation and Geoinformation*, *4*(2), 109–117.

Zito, K. S. (2021). *A Story of Drought: A Mixed Methods Approach to Analyzing Impacts on Tree Growth in the Southern Colorado Front Range* – ProQuest. https://www.proquest.com/openview/505914aa8e47d267c5298d9640f89e24/1?cbl=18750&diss=y&pq-origsite=gscholar

## APPENDIX: EARTH ENGINE CODES

```
// For_2010 data pre-processing
  var L5 = ee.ImageCollection("LANDSAT/LT05/C02/T1_TOA")
  var image_1 = ee.Image(
  L5.filterBounds(chiangmai)
  .filterDate('2010-01-01','2010-12-31')
  .sort('CLOUD_COVER')
  .first()
  );
  var image_2 = ee.Image(
  L5.filterBounds(roi2)
  .filterDate('2010-01-01','2010-12-31')
  .sort('CLOUD_COVER')
  .first()
  );
  var image_3 = ee.Image(
  L5.filterBounds(roi3)
```

```
.filterDate('2010-01-01','2010-12-31')
.sort('CLOUD_COVER')
.first()
);
var image_4 = ee.Image(
L5.filterBounds(roi4)
.filterDate('2010-01-01','2010-12-31')
.sort('CLOUD_COVER')
.first()
);
// For MOSAIC_2010
var mosaic_1 = ee.ImageCollection.fromImages([image_4, image_3, image_2,
image_1]).mosaic()
var Landsat_2010 = mosaic_1.clip(chiangmai);
Map.addLayer(Landsat_2010, {bands: ['B4', 'B3', 'B2']}, 'FCC-2010');
//Map.centerObject(chiangmai, 9);
//Map.addLayer(chiangmai, {}, 'roi');
// For NDVI_2010
var RED= Landsat_2010.select('B3');
var NIR= Landsat_2010.select('B4');
var NDVI_2010= NIR.subtract(RED).divide(NIR.add(RED));
Map.addLayer(NDVI_2010, par2, 'NDVI-2010');
// For EVI_2010
var EVI_2010 = Landsat_2010.expression('2.5*(b2-b1)/(b2+6*b1-7.5*b0+1)', {
b2: Landsat_2010.select("B4"),
b1: Landsat_2010.select("B3"),
b0: Landsat_2010.select("B1")
})
Map.addLayer(EVI_2010, par3, 'EVI-2010');
// For HEALTH_2010
var HEALTH_2010 = NDVI_2010.add(EVI_2010).divide(2);
Map.addLayer(HEALTH_2010, par4, 'HEALTH-2010');
// For_2020 data pre-processing
var L8 = ee.ImageCollection("LANDSAT/LC08/C02/T1_TOA")
var image_5 = ee.Image(
L8.filterBounds(chiangmai)
.filterDate('2020-01-01','2020-12-31')
.sort('CLOUD_COVER')
.first()
);
var image_6 = ee.Image(
L8.filterBounds(roi2)
.filterDate('2020-01-01','2020-12-31')
.sort('CLOUD_COVER')
.first()
);
```

```
var image_7 = ee.Image(
L8.filterBounds(roi3)
.filterDate('2020-01-01','2020-12-31')
.sort('CLOUD_COVER')
.first()
);
var image_8 = ee.Image(
L8.filterBounds(roi4)
.filterDate('2020-01-01','2020-12-31')
.sort('CLOUD_COVER')
.first()
);
// For MOSAIC_2020
var mosaic_2 = ee.ImageCollection.fromImages([image_8, image_7, image_6, image_5]).mosaic()
var Landsat_2020 = mosaic_2.clip(chiangmai);
Map.addLayer(Landsat_2020, {bands: ['B5', 'B4', 'B3']}, 'FCC-2020');
// For NDVI_2020
var RED= Landsat_2020.select('B4');
var NIR= Landsat_2020.select('B5');
var NDVI_2020 = NIR.subtract(RED).divide(NIR.add(RED));
Map.addLayer(NDVI_2020, par5, 'NDVI-2020');
// For EVI_2020
var EVI_2020 = Landsat_2020.expression('2.5*(b2-b1)/(b2+6*b1-7.5*b0+1)', {
b2: Landsat_2020.select("B5"),
b1: Landsat_2020.select("B4"),
b0: Landsat_2020.select("B2")
})
Map.addLayer(EVI_2020, par6, 'NDVI-2020');
// For HEALTH_2020
var HEALTH_2020 = NDVI_2020.add(EVI_2020).divide(2);
Map.addLayer(HEALTH_2020, par7, 'HEALTH-2020');
// For IMAGE_DIFFERENCE
var image_diff = HEALTH_2010.subtract(HEALTH_2020)
Map.addLayer(image_diff, par8, 'IMAGE_DIFFERENCE');
//Export
Export.image.toDrive({
image: image_diff,
description: "image_diff",
region: chiangmai,
scale: 30,
crs: "EPSG:23948"
})
```

**Change in forest cover:**

```
// For_2010
var L5 = ee.ImageCollection("LANDSAT/LT05/C02/T1_TOA")
```

```
var image_1 = ee.Image(
L5.filterBounds(chiangmai)
.filterDate('2010-01-01','2010-12-31')
.sort('CLOUD_COVER')
.first()
);
var image_2 = ee.Image(
L5.filterBounds(roi2)
.filterDate('2010-01-01','2010-12-31')
.sort('CLOUD_COVER')
.first()
);
var image_3 = ee.Image(
L5.filterBounds(roi3)
.filterDate('2010-01-01','2010-12-31')
.sort('CLOUD_COVER')
.first()
);
var image_4 = ee.Image(
L5.filterBounds(roi4)
.filterDate('2010-01-01','2010-12-31')
.sort('CLOUD_COVER')
.first()
);
// For MOSAIC_2010
var mosaic_1 = ee.ImageCollection.fromImages([image_4, image_3, image_2,
image_1]).mosaic()
var Landsat_2010 = mosaic_1.clip(chiangmai);
Map.addLayer(Landsat_2010, {bands: ['B4', 'B3', 'B2']}, 'FCC-2010');
var sample = water.merge(forest).merge(others)
var bands = ['B1','B2','B3','B4']
var input = Landsat_2010.select(bands)
var training = input.sampleRegions({
collection:sample,
properties: ['class'],
scale:30})
var Classifier = ee.Classifier.smileCart().train({
features: training,
classProperty: 'class',
inputProperties: bands})
//CLASSIFY THE INPUT IMAGERY
var L_2010 = Landsat_2010.select(bands).classify(Classifier);
//COLOUR PALETTE
var palette = [
'1618ff', // water (0) // blue {htmlcolorcodes.com}
'2ac132', // forest (1) // green
```

```
'ffb58e', // others (2) // yellow
];
//DISPLAY COLOUR RESULT
Map.addLayer(L_2010, {min:0, max:2, palette: palette}, 'LULC-2010');
// For_2020
var L8 = ee.ImageCollection("LANDSAT/LC08/C02/T1_TOA")
var image_5 = ee.Image(
L8.filterBounds(chiangmai)
.filterDate('2020-01-01','2020-12-31')
.sort('CLOUD_COVER')
.first()
);
var image_6 = ee.Image(
L8.filterBounds(roi2)
.filterDate('2020-01-01','2020-12-31')
.sort('CLOUD_COVER')
.first()
);
var image_7 = ee.Image(
L8.filterBounds(roi3)
.filterDate('2020-01-01','2020-12-31')
.sort('CLOUD_COVER')
.first()
);
var image_8 = ee.Image(
L8.filterBounds(roi4)
.filterDate('2020-01-01','2020-12-31')
.sort('CLOUD_COVER')
.first()
);
// For MOSAIC_2020
var mosaic_2 = ee.ImageCollection.fromImages([image_8, image_7, image_6,
image_5]).mosaic()
var Landsat_2020 = mosaic_2.clip(chiangmai);
Map.addLayer(Landsat_2020, {bands: ['B5', 'B4', 'B3']}, 'FCC-2020');
//Map.addLayer(Landsat_2020, parl, 'True_color')
var sample = water.merge(forest).merge(others)
var bands = ['B2','B3','B4','B5']
var input = Landsat_2020.select(bands)
var training = input.sampleRegions({
collection:sample,
properties: ['class'],
scale:30})
var Classifier = ee.Classifier.smileCart().train({
features: training,
classProperty: 'class',
```

```
inputProperties: bands})

//CLASSIFY THE INPUT IMAGERY
var L_2020 = Landsat_2020.select(bands).classify(Classifier);
//COLOUR PALETTE
var palette = [
'1618ff', // water (0) // blue {htmlcolorcodes.com}
'2ac132', // forest (1) // green
'ffb58e', // others (2) // yellow
];

//DISPLAY COLOUR RESULT
Map.addLayer(L_2020, {min:0, max:2, palette: palette}, 'LULC-2020');
//CHANGE DETECTION
var LULC_change = L_2020.subtract(L_2010);
Map.addLayer(LULC_change, {min:0, max:2, palette: palette}, 'LULC change');
```

# 7 Optimization of Solid Waste Collection and Distribution Using GIS Technology

*Subhankar Roy and Gouri Sankar Bhunia*

## 7.1 INTRODUCTION

In India, there is a serious issue with the management of municipal solid waste because of the enormous amounts that are produced daily as well as environmental and aesthetic concerns. Even though only 31% of Indians live in urban areas, the country's 377 million people (as of the 2011 Census of India) produce a staggering 143,449 metric tonnes of municipal solid waste each day, according to the Central Pollution Control Board (CPCB), 2014–15, and these numbers rise every day as the population grows. The operation of waste management entails many challenges, and among them, waste collection is the most challenging because of the significant expenses involved (Gerdes and Gunsilius, 2010). A total of 1.15 lakh metric tonnes per day (TPD) or 42.0 million tonnes of municipal solid waste are produced in urban India each year, of which 83,378 TPD are produced in 423 Class-I cities. Waste produced in 423 Class-I cities' accounts for 72.5% of all waste produced daily, and this needs to be addressed first. Municipal solid waste is made up of 40%–55% inert material, 30%–55% biodegradable (organic) matter, and 5%–15% recyclables (CPCB India, 2018). For several municipalities in developing countries, the cost of waste collection accounts for over 70% of the budget for solid waste management (Rotich et al., 2006), compared to 60% or less in developed nations (Brunner and Fellner, 2007). According to estimates, between 80% and 90% of municipal waste is disposed of in landfills without using adequate management techniques and burning it openly, which pollutes the air, water, and soil (Ahluwalia and Patel, 2018). The mobility of waste to various facilities, including transfer stations, provisional storage sites, and landfills, as well as the operating expenses and operational of this infrastructure, are all included in the total cost of solid waste management in several municipalities of cities in developing countries (Bhattacharya et al., 2018). This has compelled numerous towns, particularly in the industrialized world, to launch cost-effective research projects like route optimization (De Bercegol et al., 2017).

India has recently emerged as a market for recycling, although recycling performance has fallen short of expectations. However, a few Indian cities, like Indore, Surat, Alleppey, Bobbili, Panaji, and Pune, have demonstrated a favourable attitude towards the choice of SWM tactics. Additionally, it has been noted that towns are

 DOI: 10.1201/9781032683928-7

mostly concentrating on the collection aspect while neglecting advance treatment (Kumar and Agrawal, 2020). Swachh Bharat Mission, the government of India's flagship programme, was introduced in 2014 with the goal of providing every family with access to basic sanitation services, including toilets, as well as embracing scientific methods for the collection, processing, and disposal of municipal solid waste (Mohanty et al., 2021). The mission places a strong emphasis on each stakeholder's commitment to bringing about a discernible change in society along with the quality and sustainability of the service delivery.

Any SWM system's collection/transport component serves as its centrepiece since it makes it simple to assess the system's effectiveness and expenses. The operation includes moving garbage from production or assembly locations to transfer stations, processing facilities, or final dumping sites. As a result, it has the greatest impact and is the most expensive part of the waste management system because it consumes the largest portion of the cash municipalities have set aside for SWM.

Modern integrated GIS technology gives decision-makers access to a sophisticated modelling framework that allows them to assess and simulate a variety of SWM-related issues. In fact, the GIS tool has been used to simulate a number of waste management applications, including landfill and transfer station placement, collection and transportation optimization, and local forecasting of garbage (Karadimas and Loumos, 2008). On the basis of suitable route optimization software, a number of models for the collection and transportation of SW have been established. Additionally, the location of recycling drop-off locations has been accomplished using GIS technology (Chang and Wei, 2000); waste management in coastal areas has been optimized (Sarptas et al., 2005); solid waste generation has been estimated using local demographic and socioeconomic data (Vijay et al., 2005); and waste generation prediction at the local level (Dyson and Chang, 2005). In order to improve the effectiveness of waste collection and transportation in Patna city, Bihar (India), an optimization was created using the QGIS Network Analyst tool. This involved allocating waste bins and optimizing vehicle routing in terms of travelled distance and operational parameters while taking into account all the necessary setting parameters, including population density, waste generation rate, bins locations, and collection vehicles.

## 7.2 STUDY AREA

The capital of the state of Bihar is Patna, which is located 15 kilometres from where the Ganges and the Ganges rivers converge. It is a completely landlocked state that is situated in a transitional region with regard to climate, economy, and culture. It is situated halfway between the humid West Bengal in the east and the subhumid Uttar Pradesh in the west. On the southern bank of the Ganga River is Patna city, a significant administrative and academic hub. The Ganga, Sone, and Punpun rivers all encircle the city on three sides, giving it an extremely lengthy riverfront. The Patna Urban Agglomeration (PUA) Area has seen rapid expansion at a rate of 48.13% (1991–2001), which is higher than the average growth rates for the Patna district (30.6%) and the state (28%). This issue is a result of both internal population growth in Patna and migration from the hinterland. The Patna Municipal Corporation is

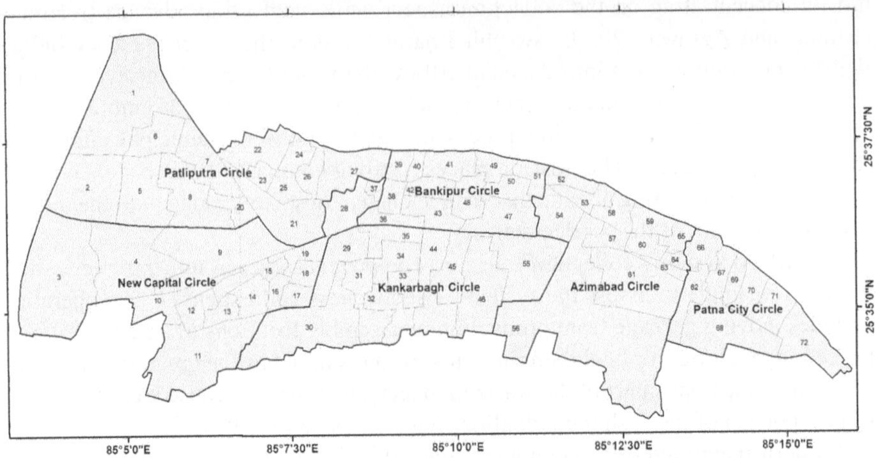


**FIGURE 7.1** Location map of the Patna municipal corporation.

in possession of 108.87 km$^2$. There are 75 wards allocated under the executive administration of six administrative circles: Azimabad Division, Bankipur Division, Kankarbagh Division, New Capital Division, Patliputra Division, and Patna City Division (Figure 7.1). As of now, 2,321,000 people are expected to live in Patna city in 2023. There are 1,234,931 people who can read and write in Patna city, 685,849 of whom are men and 549,082 of whom are women. Male and female literacy rates in Patna were 87.35% and 78.89%, respectively, with the city's average literacy rate being 83.37%. A total of 885 women for every 1000 men live in Patna city. For every 1000 boys, there are 877 girls. 13,696 slums total, with a population of 77,034 people, are located in Patna city and its outgrowth. There are 1,684,297 people living in Patna city and its expansion, or about 4.57% of the city's total population. In the city of Patna, 86.39% of the population practise Hinduism. With 12.27% of Patna residents practising it, Islam is the second most prevalent religion. Buddhism comes in at 0.02%, Jainism at 0.09%, Sikhism at 0.23%, and Christianity at 0.51% in Patna city.

## 7.3 MATERIALS AND METHODS

### 7.3.1 Data Collection and Database Creation

Utilizing the GIS environment "Quantum Geographic Information System" (QGIS), a spatial database was created using maps, data from municipal and statistics agencies, Open Street Map (OSM), monitoring and fieldwork, and information from literature. The necessary information relates to the research area's geographic and urban features as well as the specifics of the garbage collection process. The ward and circle map have been georeferenced and digitized from the Global Positioning Data using QGIS software. The following information was gathered and processed in the appropriate formats (vectors, tables, and raster): the delineation of the study site, the municipality's detailed land use plan, the population distribution and density, a satellite image (Google Earth), the road network, and the current waste collection infrastructural facilities. Ward-wise population data was collected from the municipal office. All necessary information and facts have been gathered, including the starting point's location, the number of workers, the route taken and the coordinates of the collection points, the state of the waste when it was generated, the state of the containers and bins, the transfer station, the volume of waste that was collected, and the amount of energy used.

### 7.3.2 Heatmap of GVP

We sequentially collected GVP data from each ward of Patna Municipal Corporation. In this study, a total of 483 GVP locations were identified and geographic location of each of the GVP was recorded through online GPS Map Camera mobile app. Finally, all the points recorded on GIS platform and shapefile was created of GVP location within the city. Heatmap of GVP locations were analysed based on QGIS software, with 10-millimetre radius. The quantity of points in a location is used to compute the density. When there are fewer GVPs, the area they cover within the municipality is larger and they are less dense; in contrast, when there are more GVPs, there are more GVP sites inside the cluster.

### 7.3.3 Route Optimization

Firstly, it is considered that the nodes formed at street intersections serve as potential secondary collection places for a given road network. Hubness seems to be associated with the occurrence of distance concentration, which is usually expressed as the proportion between some criterion of dispersion (e.g., standard deviation) and some assess of magnitude (e.g., the mean) of distances of all GVP points in a data set to some arbitrary reference point (Aggarwal et al., 2001; Franc'ois, 2007). Distance to nearest hub tool creates lines that join each feature of an input garbage vulnerable points (GVPs) to the nearest point in a secondary collection point. Distances are calculated based on the centre of each secondary collection point. By using the sample mean of the GVP distribution as a point of comparison, it can be seen that a GVP point's location in the data space significantly affects the value of its $k$ occurrences. For combinations of secondary collection point distributions and distance measures for which hubness occurs, as well as for various values of $k$, we also made equivalent observations. Secondary collection hubs typically look near to the means of the individual GVP distributions of multimodal data distributions, such as those formed from a combination of unimodal distributions. The solution is then determined by a function that considers a number of factors, including the shortest distance, secondary collection points, GVP locations, and the road network.

### 7.3.4 Statistical Analysis

Descriptive statistics of the waste collection characteristics have been done through Microsoft Excel. Cross-correlation analysis is performed to estimate the relationship between infrastructure availability and waste production characteristics in the study area. All the analysis is performed at 0.05 significance level.

## 7.4 RESULTS

The development of a solid waste management system requires evaluation of trash generation. The overall weight and volume of trash generated by the community the management system serves are more crucial in terms of planning and design, even though per capita waste output is a metric that is required for demonstrating trends in consumption and production. Figure 7.2 illustrates the spatial distribution of population density in Patna Municipal Corporation. Based on the population density, the study area is divided into four categories, namely (i) <1500 persons per sq km, (ii) 1500–3000 persons per sq km, (iii) 3000–4500 persons per sq km, and (iv) 4500 persons per sq km. Results show the maximum population density is distributed in the northern part, north-east, and small pockets of central region in Patna municipality. The lowest population density is distributed in the southern and western parts of the municipality. The medium population density is distributed in the central part of the study area.

Figure 7.3 illustrates the spatial density of GVPs in Patna Municipal Corporation. The highest density of GVPs are mainly distributed in the Kankarbagh and Bankipur circles. The lowest GVPs density are mainly observed in the Azimabad and New Capital circles. The lowest GVPs density is observed in Patna and Patliputra circles.

The amount of garbage created will also change as a result of the growing population, which is directly related to changing lifestyles and consumption patterns. Urban inhabitants' daily lives provide evidence of the shifting consumption patterns, such as their propensity to purchase fast food that uses, for instance, throwaway food containers. The growth in waste creation is significantly impacted by this consumption style. The city of Patna serves as a major draw for migrant populations. The population is growing due to migration as well as natural birth rates.

The growing generation of municipal solid trash is seen to be a result of population density, which is seen as a threat to the environment. There are various restrictions

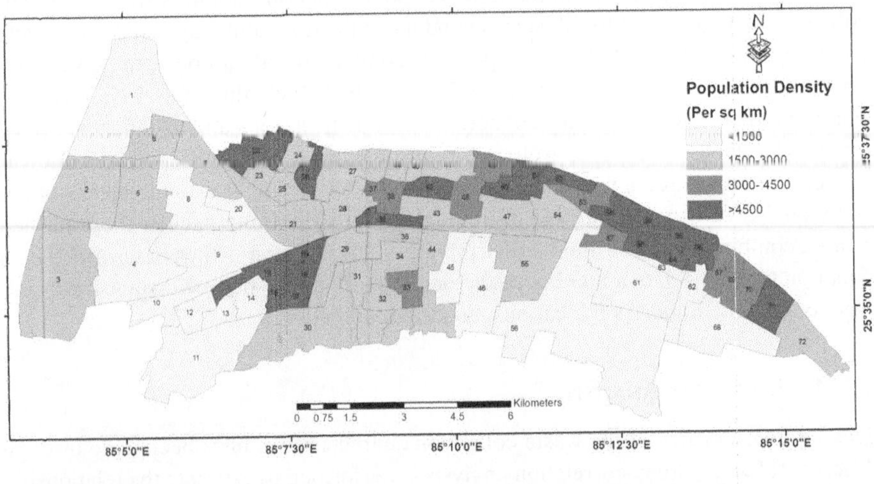


**FIGURE 7.2** Spatial distribution of ward-wise population density in Patna municipal corporation.

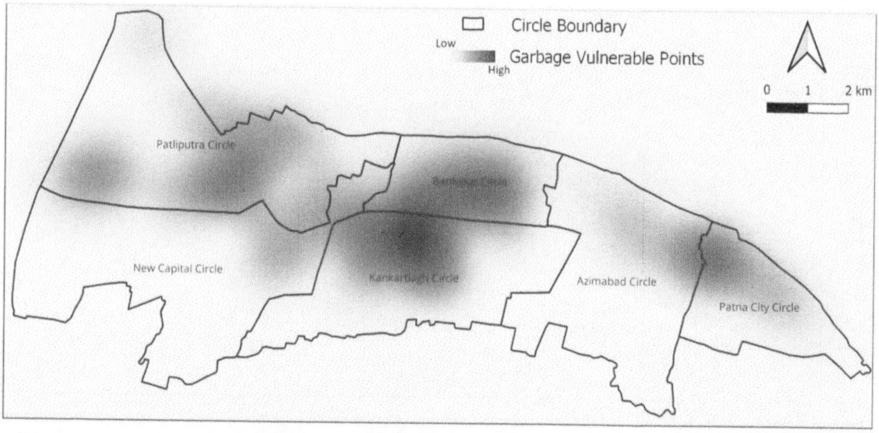


**FIGURE 7.3** Heatmap of garbage vulnerable points in Patna municipal corporation.

on waste handling, both in terms of transportation and issues with final shelter, as a consequence of the population density and the range of life levels in the population. Figure 7.4a illustrates amount of waste generation and its geographical area. Results showed moderate and positive relation ($R^2 = 0.37$) in the study area. Figure 7.4b showed the correlation between population distribution and the amount of waste generation in Patna Municipal Corporation. Results showed there is strong positive correlation ($R^2 = 0.93$) between population and waste generation in the study area. A significance p-value of less than 0.05 confirms this finding.

A cross-correlation matrix is created to identify the relationship between waste management assets and estimated waste generation in the study area. Results showed moderate positive relationship between number of vehicles and estimated waste generation per sq km ($r^2 = 0.319$) and waste generation per lakh population ($r^2 = 0.420$) in the study area. Moreover, a strong positive correlation is observed between workforce and waste production/in TPD ($r^2 = 0.849$) and total waste generation ($r^2 =$ 0.844). The strong negative relationship is observed between waste generation per sq km ($r^2 = -0.690$) and waste generation per lakh population ($r^2 = -0.616$).

Figure 7.5 represents the circle-wise spatial variation of waste management practices adopted by Patna Municipal Corporation. Figure 7.5a showed the spatial distribution of workforce in which the maximum number of workforces is allocated for Azimabad Circle and the lowest number of workforces is allocated for Bankipur Circle. Figure 7.5b showed the maximum number of vehicles is allocated for New Capital Circle followed by Bankipur Circle. The lowest number is assigned for Kankarbagh Circle. Figure 7.5c showed the spatial distribution of quantity of waste generation in PMC.

Road network map has been prepared which covers all the possible streets requiring services. An attribute of road class is added as per the survey map of the study area (Figure 7.6). The command area of a secondary collection point, contributing solid waste to that particular location is computed considering the shortest distance.

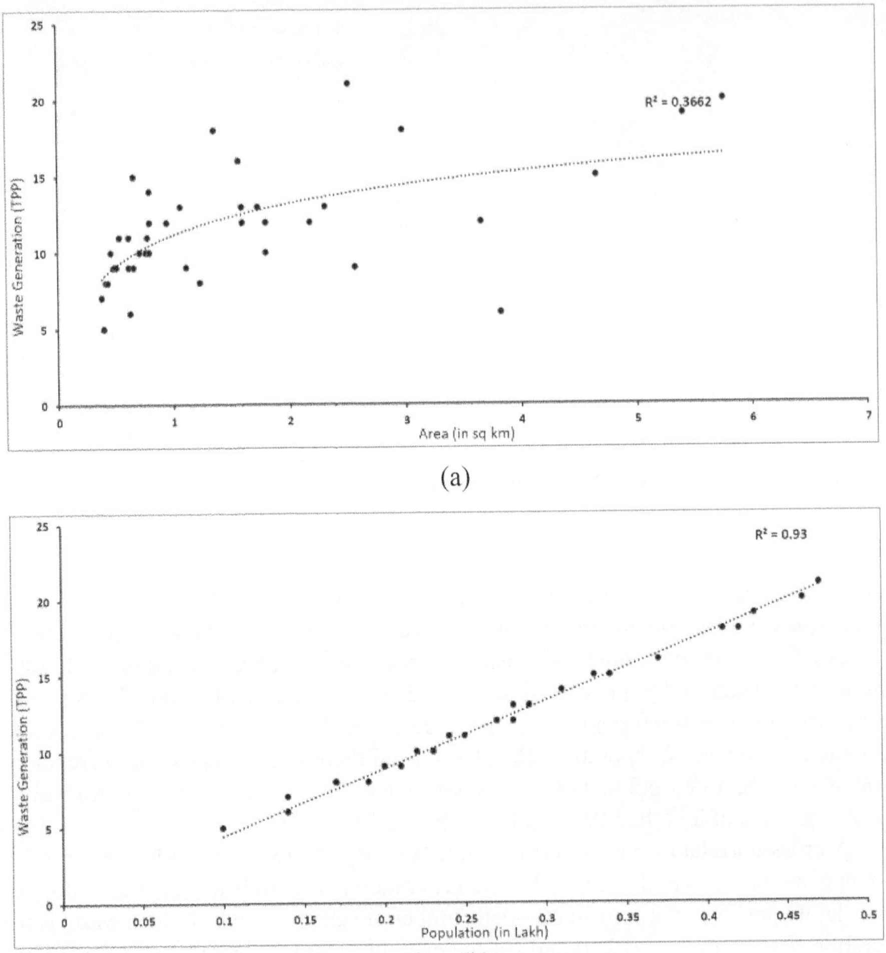


(b)

**FIGURE 7.4** Correlation between waste generation and area in sq km (a) and number of population (b) in Patna municipal corporation.

The connectivity between the nearby secondary collection locations acts as the network's links, while the centres of each secondary collection point act as its nodes. Based on the distance to hub tool used for the new route identification with the help of existing road network and location of secondary collection points to identify the shortest route which will cover less distance and covers maximum waste collection points with a less overlap. Figure 7.6 suggested the circle-wise newly shortest route and connectivity among the GVPs and secondary collection points in the study area. The revised route decreased the overall distance travelled, resulting in a gradual reduction in fuel usage. As a result, the expenses associated with collection (collecting) and pollution emissions decreased.

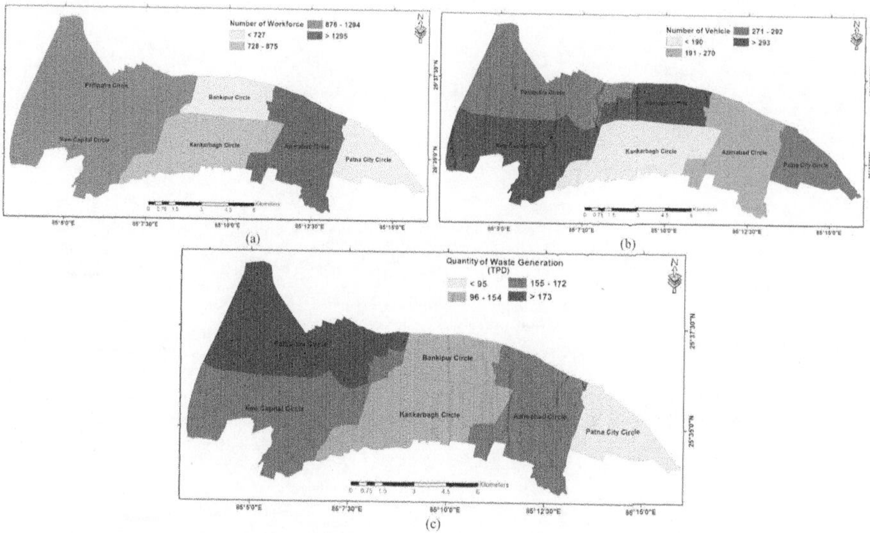


**FIGURE 7.5** Circle-wise (a) workforce distribution, (b) vehicle allotment, and (c) quantity of waste generation in Patna municipal corporation.

Table 7.1 displays the distance for both the current and ideal garbage collection routes for automobiles. Table 7.1 displays the precise distance-based percentage changes for all parcels between the current and optimized routes. For the collection and delivery of waste, route optimization studies often concentrate on determining the shortest distance or minimal driving durations. The maximum difference of route is calculated for New Capital Circle (12.76 km), followed by Kankarbagh Circle (10.47 km). The lowest difference is calculated for Patna City Circle (5.78 km), followed by Bankipur Circle (6.04 km). Moreover, the maximum saving efficiency is calculated for Bankipur Circle (31.09%), followed by Patna city (26.06%). The lowest saving efficiency is calculated for Patliputra (15.05%), followed by New Capital (17.30%) Circle.

## 7.5 DISCUSSION

Due to the adoption of a contemporary lifestyle, the ever-growing population in Patna Municipal Corporation has caused the generation of solid waste, resulting in the degradation of the land, water, and air resources. Household solid wastes—also referred to as trash or garbage—are the solid waste produced by various urban local entities (Sinchana et al., 2021). A growing number of instances involving risks to human health have been caused by the unregulated and improper disposal of such garbage. In the management of solid waste and garbage collection places, GIS can be used successfully (Wu et al., 2020). Secondary collection stations must be assigned by determining proximity distances that are convenient to the people in order to prevent open dumps of waste. To prevent bin overflow, bin size should be chosen taking into account the quantity of waste produced and the frequency of collection.

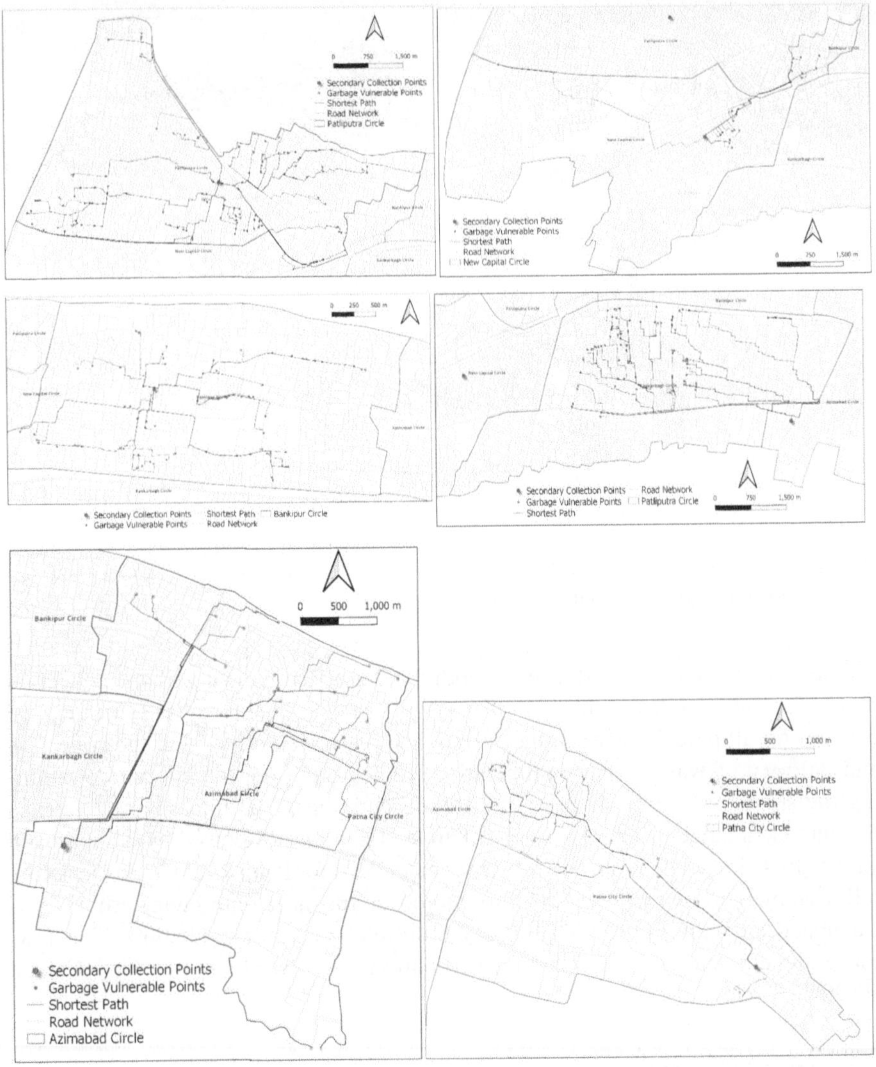


**FIGURE 7.6** Circle-wise shortest path analysis between secondary collection points and garbage vulnerable points.

Waste collection routes may be properly decided using GIS, which will save collection costs. Also, this approach aids in identifying the intermediary nodes from which the shortest path is determined. User can easily determine the path from which the shortest distance can be acquired.

The QGIS tool was used in this study to build an optimization in order to increase the effectiveness of waste collection and transportation in the Patna municipality of Bihar. Large-scale MSW production in the Patna metropolitan regions has turned into a significant environmental problem. Urban municipal governments are devoted

**TABLE 7.1**
**Circle-wise Comparison between Existing Route and Optimize Route**

| Circle | Existing Route | Optimize Route | Difference (in km) | Saving Efficiency (%) |
|---|---|---|---|---|
| Bankipur | 19.43 | 13.39 | 6.04 | 31.09 |
| Patna City | 22.18 | 16.4 | 5.78 | 26.06 |
| Azimabad | 26.54 | 19.66 | 6.88 | 25.92 |
| Patliputra | 50.62 | 43 | 7.62 | 15.05 |
| New Capital | 73.76 | 61 | 12.76 | 17.30 |
| Kankarbagh | 56.21 | 45.74 | 10.47 | 18.63 |

to providing services, but they are having trouble managing this situation effectively due to the increased severity of issues. The main issues with MSW management at PMC are caused by inadequate waste segregation at the source, a low level of house-to-house collection, a large number of open vats, ineffective waste transport systems with outdated vehicles, ineffective collection in newly acquired areas, and an ineffective informal recycling system.

Network analysis is used in this study to determine the best path for collecting waste from different waste collection points throughout the city (Malakahmada et al., 2014). The influence of the weather and road conditions on driving, however, was not examined in this study. From each pickup point to the closest garbage site, we calculated the shortest or best path. According to this study, employing efficient routing and procedures can result in significant cost savings. This can be done by cutting down on overall route distance. According to this study, employing efficient routing and procedures can result in significant cost savings. This can be done by cutting down on overall journey distance.

## REFERENCES

Aggarwal, R. K., Singh, J., & Garg, V. K. (2001). Municipal solid waste management in Indian cities: A review. Indian Journal of Environmental Protection, 21(6): 521 528.

Ahluwalia I.J., Patel U. 2018. Working Paper No. 356 Solid Waste Management in India An Assessment of Resource Recovery and Environmental Impact Isher Judge Ahluwalia. Indian Council for Research on International Economic Relations (Issue 356). https://icrier.org/pdf/Working_Paper_356.pdf

Bhattacharya R.R.N.S., Chandrasekhar K., Deepthi M.V., Roy P., Khan A. 2018. Challenges and Opportunities: Plastic Waste Management in India. The Energy and Resources Institute, 24(2). Available at: https://www.teriin.org/sites/default/files/2018-06/plastic-waste-management_0.pdf

Brunner, P. H., & Fellner, J. 2007. Setting priorities for waste management strategies in developing countries. Waste Management & Research, 25(3): 234–240. https://doi.org/10.1177/0734242X07078296

Chang N., Wei Y.L. 2000. Siting recycling drop-off stations in urban area by genetic algorithm-based fuzzy multi objective nonlinear integer programming modelling. Fuzzy Sets and Systems, 14 133 149 0165-0114

CPCB India, 2018. MSW- Generation & Composition CPCB 2018. Available at: https://cpcb.nic.in/uploads/MSW/Waste_generation_Composition.pdf

De Bercegol R., Cavé J., Nguyen Thai Huyen A. 2017. Waste Municipal Service and Informal Recycling Sector in Fast-Growing Asian Cities: Co-Existence, Opposition or Integration? Resources, 6 (4): 70, DOI: 10.3390/resources6040070

Dyson B., Chang N.. 2005 Forecasting municipal solid waste generation in a fast-growing urban region with system dynamics modelling, Waste Management, 25: 669 679 0095-6053X

Franc'ois D. 2007. High-dimensional Data Analysis: Optimal Metrics and Feature Selection. PhD thesis, Universite catholique de Louvain, Louvain, Belgium.

Gerdes P., Gunsilius E. 2010. The Waste Experts: Enabling Conditions for Informal Sector Integration in Solid Waste Management. In Management, Available at: www.gtz.de/recycling-partnerships

Karadimas N.V., Loumos V.G. 2008 GIS-based modeling for the estimation of municipal solid waste generation and collection. Waste Management & Research 26 337 346 1096-3669

Kumar A., Agrawal A., 2020. Recent Trends in Solid Waste Management Status, Challenges, and Potential for the Future Indian Cities – A Review. Current Research in Environmental Sustainability, 2.

Malakahmada A., Bakria P.M., Radin M., Mokhtara K.N. 2014. Solid Waste Collection Routes Optimization via GIS Techniques in Ipoh City, Malaysia. Fourth International Symposium on Infrastructure Engineering in Developing Countries, IEDC, Procedia Engineering, 77: 20–27.

Mohanty S., Saha S., Santra G.H., Kumari A. 2021. Future Perspective of Solid Waste Management Strategy in India. In: Baskar, C., Ramakrishna, S., Baskar, S., Sharma, R., Chinnappan, A., Sehrawat, R. (eds) Handbook of Solid Waste Management. Springer, Singapore. https://doi.org/10.1007/978-981-15-7525-9_10-1

Rotich, K. H., Zhao, Y., & Dong, J. 2006. Municipal solid waste management challenges in developing countries – Kenyan case study. Waste Management, 26(1): 92–100. https://doi.org/10.1016/j.wasman.2005.03.007

Sarptas H., Alpaslan M.N., Dolgen D. 2005. GIS supported solid waste management in coastal areas. Water Science and Technology, 51(11): 213–220.

Sinchana N.L., Apoorva K.S., Mutturaj, Niranjan P.T. 2021. Route Optimization for Effective Municipal Solid Waste Collection System in Selected Wards of Mysuru City. International Research Journal of Engineering and Technology, 8(7): 1381–1384

Vijay R., Gupta A., Kalamdhad A.S., Devotta S. 2005 Waste Management & Research Estimation and allocation of solid waste to bin through geographical information systems. 0073-4242 5 23 479 484

Wu H., Tao F., Yang B. 2020. Optimization of Vehicle Routing for Waste Collection and Transportation. International Journal of Environmental Research and Public Health, 17(14): 4963. https://doi.org/10.3390/ijerph17144963

# 8 Spatial Association between Population Characteristics and Visceral Leishmaniasis Incidence in Muzaffarpur District, Bihar (India)

*Saroj Senapati, Manju Pandey, and Gouri Sankar Bhunia*

## 8.1 INTRODUCTION

Visceral leishmaniasis (VL) or kala-azar is a poverty-related infection, which affects the poorest among the poor. Previous studies illustrated that malnutrition, displacement, poor housing, illiteracy, gender discrimination and sickness of the immune system and deficit of resources are typically associated with the occurrence of VL (Desjeux, 2001; Thakur, 2000; http://www.emro.who.int). The characteristics of a population can help to determine the possible impact of VL infection and the trends and patterns of VL over space and time. The association between population characteristics and the VL infection is difficult to measure; however, previous study showed certain factors do affect a person's exposure to VL infection (Actor, 1960; Cerf et al., 1987; Dye and Williams, 1993). Some risk factors are clearly man made, such as migration, deforestation, urbanization or changes in the human host's susceptibility to infection, such as immunosuppression and malnutrition (Assimina et al., 2008; Roland, 2002). Alternatively, socioeconomic factors are conditions that may affect the VL transmission and propagation of a person or group. These factors include education, occupation and income and related to VL infection. Preliminary studies conducted by Kesari et al. (2010), Costa et al. (2005) and World Health Organization (WHO) (2002) showed households and demographic characteristics are intimately associated with VL transmission. These kinds of associations are important to consider when deciding whether connections exist between the population characteristics and VL outcomes.

In VL endemic areas, augmented infection risk is interceded through poor housing structures and environmental cleanliness, need of personnel fortification measures and economically obsessed migration and employment that fetch people in to

DOI: 10.1201/9781032683928-8

get in touch with infected sand flies (Alvar et al., 2006). The importance of VL as a hindrance to economic development is still not fully recognized although there is little doubt that the disease is poverty-related (Alvar et al., 2006). Bihar in India, one of the economically most deprived states where 80% of the families in VL-affected villages were found to belong to the poorest two quantities in terms of wealth distribution (Boelaert et al., 2009). Over the last decade there have been at least seven attempts to estimate the financial and economic costs incurred by VL cases in India (Meheus et al., 2006; Sarnoff et al., 2010; Sundar et al., 2010), Nepal (Adhikari et al., 2009; Rijal et al., 2006; Sharma et al., 2004) or Bangladesh (Sharma et al., 2006). The principal cost driver was observed to be the income lost because of illness (i.e. an indirect cost), which may correspond to 60% of the total household expenditure (Meheus et al., 2006).

In the present study, considering the socioeconomic characteristics of the population, a quantitative and qualitative assessment was carried out to measure the 'high' and 'low' kala-azar transmission areas. The village-level population characteristics data, such as population density, illiteracy rate, unemployed population, family size, agricultural density and nutritional density was collected from the census and remote sensing data to demarcate the 'high' and 'low' kala-azar transmission zone based on Geographical Information System (GIS) platform.

## 8.2 MATERIALS AND METHODS

### 8.2.1 Study Area

Muzaffarpur district is situated on the northern Bihar in India, extended from 25°54′ 33.83″N to 26°22′57.38″N latitude and 84°52′17.07″E to 85°45′51.34″E longitude (Figure 8.1). The district had a population of 4,778,610 of which male and female populations were 52.68% and 47.32%, respectively (as per 2011 Census, India, http://www.census2011.co.in/census-/district/68muzaffarpur.html). The population density of the district is 1,506 persons km$^2$. In Muzaffarpur District, the total literate populations were 2,601,665 of which male and female were 59.16% and 40.84%, respectively. The average literacy rate of Muzaffarpur was higher in 2011 (65.68%) compared to 2001 (47.95%). The total working population is 37.93% of the total population, while 58.19% of the men and 15.86% of the women are working population, respectively. The main working population is 29.81%, while 62.07% is non-working population of the total population in the study area.

### 8.2.2 Census Data Collection and Base Layer Preparation

In this study, location data and census data for Muzaffarpur district at the village level were collected from the Department of Census, Bihar, India (Census of India, 2001, http://censusindia.gov.in/). In this research, location data and population data for the 2,047 villages of Muzaffarpur districts were collected from the census records (Available at: http://muzaffarpur.bih.nic.in/). Annual population estimates for each census tract were calculated by the projected population growth equation

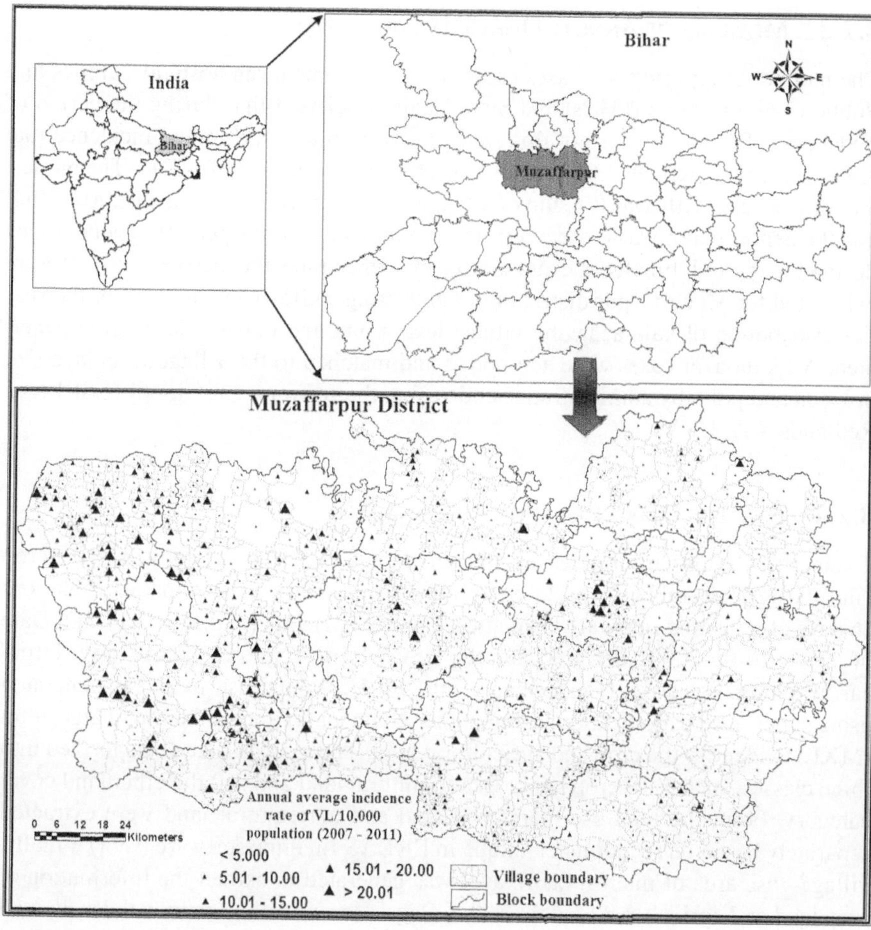


**FIGURE 8.1** Location map of the study area.

(Equation 8.1) with an annual component of population data from the 2001 demographic census.

$$\textbf{\textit{Projected population}} = \textit{Population of the demographic census year} \times \left(1+\frac{2.38}{100}\right)^{\wedge}\left(\textit{Population of the estimated year} - 1\right) \quad (8.1)$$

The indicators were developed to reflect some key characteristics in relation to the VL transmission socioeconomic condition and population in the census tracts in the study period, namely (i) population density, (ii) illiteracy rate, (iii) family size, (iv) percentage of the unemployed population, (v) agricultural density and (vi) nutritional density.

### 8.2.3 Measures of Annual Disease Incidence Rate

The number of reported VL cases came from government-ran hospitals, clinics and Public Health Centers (PHCs) and State Health Society, Bihar during the period of 2007–2011. For the purpose of this study, a five year average annual incidence rate (CIR) was calculated per 1,000 populations for the period of 2007–2011. The numerator of five years CIR was the sum of all kala-azar cases recorded during 2007–2011, and the denominator was the average number of people at risk per 1,000 populations during this period. Finally, the cumulative incidence rates for each census tract were calculated for Muzaffarpur district. For conducting a GIS-based analysis of the spatial distribution of kala-azar, the village-level point and polygon layer were generated. All kala-azar cases were geo-coded and matched to the village-level layers of polygon and point by administrative code using the software ArcGIS 9.1 (ESRI Inc., Redlands, CA, USA).

### 8.2.4 Satellite Data Derivation and Analysis

To measure the agricultural density and nutritional density in the study area, the crop land and agricultural land was delineated from the Landsat 5 (30 m spatial resolution/16 days of revisiting interval) Thematic Mapper (TM) data (Path/row – 141/42; Date of pass – 16-06-2009), derived from the USGS Earth Explorer Community (http://earthexplorer.usgs.gov/) (Herbreteau et al., 2005). A land use map was generated using supervised classification technique based on a maximum likelihood algorithm (MXL) to identify the crop and agricultural land. The map has been categorized into three classes, such as (i) crop land, (ii) agricultural land and (iii) the other land cover category. From the land use map, crop land and agricultural land were extracted separately based on recoding technique in ERDAS Imagine software (v.9.1). Finally, village-wise area of interest (.aoi) layer was generated to extract the information of crop land and agricultural land for each village. However, the nutritional density and agricultural density of the study villages were calculated using the formula followed by Trewartha (1953) and Singh and Dhillon (2004) and integrated with the GIS database for further analysis.

### 8.2.5 Spatial Autocorrelation and Smoothing

To describe the spatial autocorrelation of the population parameters, the univariate global Moran's *I* statistics were calculated to comprehend whether adjacent villages having greater similarity in terms of the target indicator than would be expected considering a clustered/random pattern. The Moran's *I* statistics can take values ranging from −1 to +1 and is positive for direct correlation and negative when inverse (Câmara et al., 2002; de Almeida et al., 2011). Moran's *I* measures whether concurrent areas have a stronger resemblance to the studied indicator than that expected in a random distribution.

However, a continuous surface was derived from discrete samples with measured values using the inverse distance weighting (IDW) method. The goal of spatial interpolation is to create a surface that is intended to best represent empirical reality thus the method selected must be assessed for accuracy for local scale studies. However, the IDW interpolation technique estimates cell values by averaging the values of village data points in the neighbourhood of each processing cell (Willmott, 1984) and is sensitive to outliers. The closer a point is in the centre of the cell being approximated, the more weight, or influence it has in the averaging method. This process presumes that the parameters being diagrammed decrease in influence with distance from its sampled location (Fisher et al., 1987; Hu et al., 2007; Zarina et al., 2010), and best with evenly distributed spatial features. However, by using this procedure smoothed polygon layers were generated that yield a less heterogeneous thematic layer, usually referred to as smooth. Each of the indicators of population characteristics is categorized into a definite class based on the geometric interval.

### 8.2.6 Kala-azar Transmission Risk Model

The weighted linear combination model was used by assuming the socioeconomic characteristics such as population density, family size, unemployed population, illiteracy rate, agricultural density and nutritional density of the area either individually or in combination are known to be associated with the occurrence of kala-azar in the Indian sub-continent. Weighted linear combination, a familiar method, is applied for the common measurement scale of values to assorted and unalike inputs to produce an integrated analysis (Banai-Kashani, 1989; Malczewski, 2000; Satty, 1980). The association between population characteristics and the average annual incidence rate of VL per village-wise was tested by the Spearman rank correlation (*rho*) method to estimate their influence on VL transmission in this particular region. The analysis was executed with <0.05%.

According to the relative importance of each criteria for kala-azar transmission, simple weightings/ratings were evaluated for all of the input variables, leading to the weighted linear model. We utilized rating systems based on values from 1 to 5, where '5' means very highly suitable, '4' highly suitable, '3' moderately suitable, '2' less suitable and '1' very less/unsuitable. The model customized the environmental and socioeconomic factors encompassing the ranges of different variables associated with the transmission of kala-azar, including high population density, large family size, higher illiteracy rate, large number of unemployed population, agricultural density and high nutritional density. Data for each criterion are standardized through linear transformation (Equation 8.2).

$$S_{i=\frac{X_i - X_{min}}{X_{max} - X_{min}},} \tag{8.2}$$

where $S_i$ is the standardized value for the original value $X_i$, $X_{min}$ is the lowest original value and $X_{max}$ is the highest original value. However, the scores and weights of the each variable were assigned according to the spatial association of VL incidence rate with each category of demographic variables, estimated through overlay analysis as well as univariate statistical analysis.

Conversely, it is not possible to use Equation 8.2 because of the variables (i.e. unemployed population) of the model is nominal. Therefore, a ranking procedure based on expertise and knowledge the data converted into standardizing range such as, 1–5. Finally, each input raster is weighted a per cent influence based on its importance to the model of *rho* values. The total influence for all rasters equals 100%. The cell values of each input raster are then multiplied by the raster's weight (Equation 8.3).

$$I = \frac{\sum_{i=1}^{n} w_i x_i}{\sum_{i=1}^{n} w_i}, \tag{8.3}$$

where $I$ is the index value, $n$ is the number of criteria, $w_i$ is the weight for criteria $i$, and $x_i$ is the standardized value for criteria $i$.

In this weighted linear combination model, population density per km² has a 10% influence, family size at 30% influence, illiteracy at 25% influence, the unemployed population at 15% influence, agricultural density at 10% influence and nutritional density at 10% influence has been assigned.

## 8.3 RESULTS

### 8.3.1 Disease Incidence of the Study Area

A total of 15,905 cases and 127 deaths were recorded in the study area during the period of 2007–2011. The highest number of cases (4,920) was recorded in 2007; however, the numbers of cases were decreasing in trend (Figure 8.2). The highest incidence rate of VL was also calculated in 2007 (11.40 kala-azar cases per 10,000 populations) and the lowest incidence rate was calculated in 2011 (4.99 kala-azar cases per 10,000 populations). The mean annual incidence rate is 4.00 per 10,000 populations among the villages in Muzaffarpur district, which is recorded in (Table 8.1).

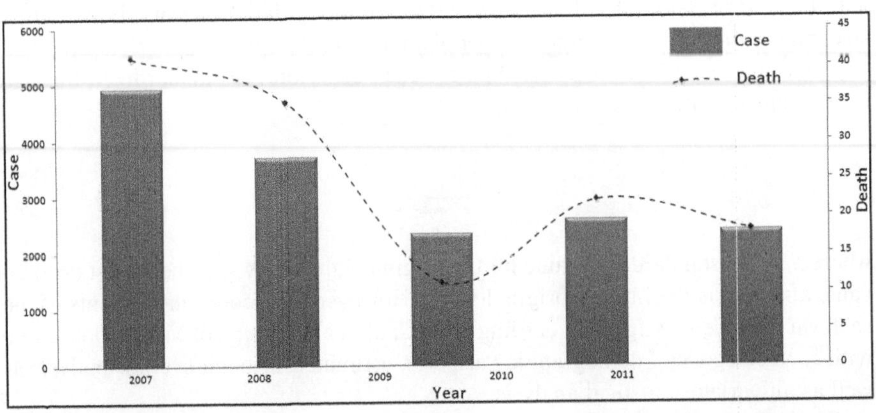


**FIGURE 8.2** Year-wise distribution of kala-azar cases and death in Muzaffarpur district.

**TABLE 8.1**
**Population Characteristics of Muzaffarpur District**

| Variables | Mean | Median | Mode | Standard Error | Standard Deviation | Kurtosis | Skewness |
|---|---|---|---|---|---|---|---|
| Average annual incidence rate/1,000 population | 4.00 | 2.33 | 1.69 | 0.14 | 5.00 | 17.24 | 3.46 |
| Family size | 5.46 | 6.05 | 6.05 | 0.03 | 1.42 | 0.37 | −0.52 |
| Illiteracy rate (%) | 64.67 | 64.99 | 100 | 0.28 | 12.39 | 0.68 | −0.14 |
| Population density (km$^2$) | 385.71 | 119.62 | 115.6 | 34.71 | 1531.39 | 175.49 | 11.63 |
| Income (%) | 43.97 | 43.51 | 43.10 | 0.23 | 10.31 | 4.77 | 0.66 |
| Agricultural density | 0.80 | 0.47 | 0.45 | 0.04 | 1.17 | 19.35 | 3.97 |
| Nutritional density | 0.09 | 0.06 | 0.01 | 0.005 | 0.19 | 14.58 | 10.71 |

### 8.3.2 Association between Population Density and Average Annual Incidence of VL

A village-level database of population density of Muzaffarpur district was generated based on the census report. The result showed that the population density of Muzaffarpur district varies from 41.20 to 1173.85 per km$^2$ (Table 8.1). Highest density was recorded in the western part of the district, and some small pockets were also delineated on the southern part and the northern part of the district. In the district, the lowest density was recorded from eastern and north-central part of the study area. Spatial autocorrelation of population density per km$^2$ among the adjacent villages in the district showed clustered pattern (*Moran's* $I = 0.36$, Z score 41.17, $p < 0.01$). Figure 8.3a also shows the relationship between population density and average annual incidence rate of VL during the period of 2007–2011 in Muzaffarpur district. It is observed that that the intensity of occurrence increases with the increase in population density in the study area. A quantitative analysis of the distribution of VL in relation to population density reveals a positive correlation (*rho* = 0.27, $p < 0.005$).

### 8.3.3 Association between Family Size and Average Annual Incidence of VL

In Muzaffarpur district, a positive and strong relationship was established between the family size and VL occurrences (*rho* = 0.47, $p < 0.000$). In the district, very small part of the area showed a family size with less than 5.0, observed in the central and northeastern part (Figure 8.3b). The spatial autocorrelation of family size among the villages across the study site showed a clustered pattern (*Moran's* $I = 0.48$, Z score 57.22, $p < 0.01$). However, the concentration of disease is also higher with the area extended by larger family size. Family size in this region varies from 2.02 to 10.41, with an average mean size of 5.46. It is, therefore, concluded that the occurrence of VL increases with the increase of family size.

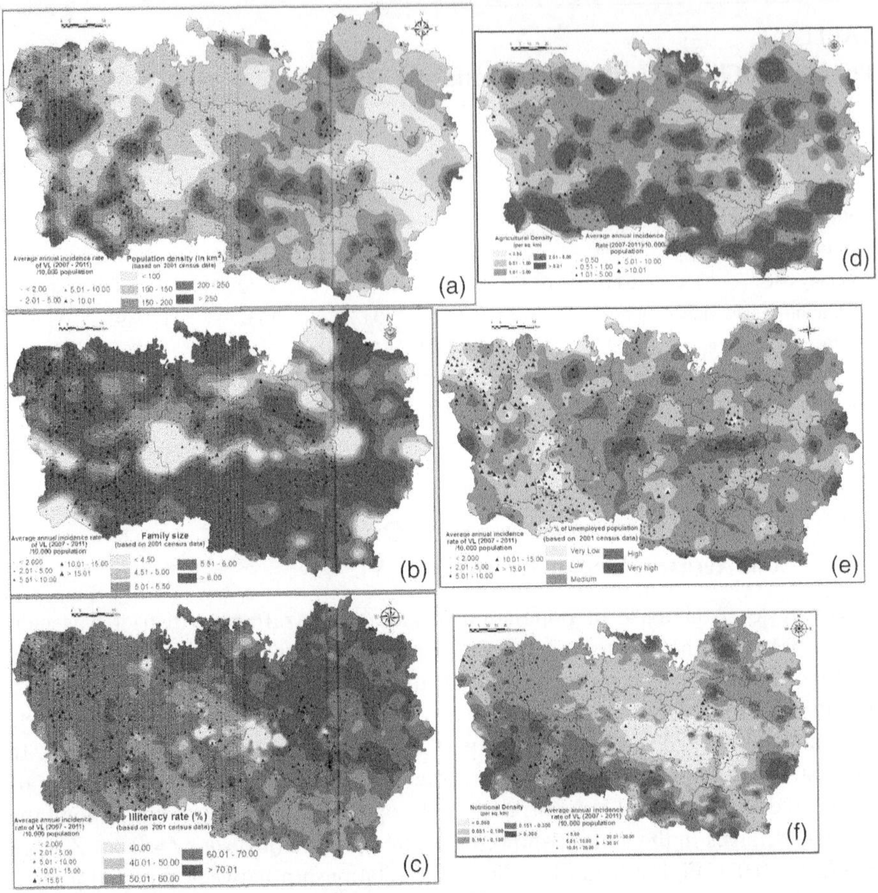


**FIGURE 8.3** Population characteristics of Muzaffarpur district (a) Population density of vs. average annual incidence of VL during the period from 2007 to 2011, (b) Family size vs. average annual incidence of VL during the period from 2007 to 2011, (c) Illiteracy rate vs. average annual incidence of VL during the period from 2007 to 2011, (d) Agricultural density vs. average annual incidence of VL during the period from 2007 to 2011, (e) Unemployed per cent of population vs. average annual incidence of VL during the period from 2007 to 2011, (f) Nutritional density vs. average annual incidence of VL during the period from 2007 to 2011.

### 8.3.4 Association between Illiteracy Rate and Average Annual Incidence of VL

The illiteracy rate of Muzaffarpur district varies from 12.16% to 87.65%, with an average of 64.67%. The lowest illiteracy rate was found in the central part of the district (less than 40%). In the study area, the highest illiteracy rate was viewed on the northern and northeastern part of the district. Some small pockets of higher illiteracy were also portrayed in western and southern part of the district (Figure 8.3c).

However, the spatial distribution of the illiteracy group of people across the district showed clustered pattern (*Moran's I* = 0.73, Z score 86.47, $p < 0.01$). Moreover, the frequency of occurrence of VL in the study area reveals positive and significant relationship (*rho* =0.38, $p < 0.05$) with the illiteracy rate, indicating that these two variables tend to increase together.

### 8.3.5 Association between Unemployed Population and Average Annual Incidence of VL

The spatial distribution of the unemployed population in Muzaffarpur district showed clustered pattern (*Moran's I* = 0.67, Z score 79.64, $p < 0.01$). Figure 8.3e shows that the per cent of unemployed population is highest in the western part of the district; some small pockets are also found in the eastern and northern part. The percentage across the district is varied from 17.3 to 98.73. The overlay analysis indicated that when the per cent of the unemployed population increase, the VL occurrence rate is also increasing. A statistical analysis by the Spearman rank correlation coefficient method confirmed the strong positive relationship between these two variables. A correlation coefficient (*rho*) of 0.37 with $p < 0.001$ was found, indicating that the two variables tend to increase together.

### 8.3.6 Association between Agricultural Density and Average Annual Incidence of VL

The agricultural density of Muzaffarpur district varies from 0.01 to 9.28 (SD±1.17). The spatial autocorrelation of the agricultural density villages showed clustered pattern (*Moran's I* = 0.51, Z score 37.26, $p < 0.01$) across the district. Based on the agricultural density values, the district is divided into the five categories using geometric interval, like (1) less than 0.5 – very low agricultural density, (2) 0.51–1.00 – low agricultural density, (3) 1.01–2.00 – moderate agricultural density (4) 2.01–5.00 – high agricultural density and (5) more than 5.01 – very high agricultural density region (Figure 8.3d). The high agricultural density region was delineated in the southern part of the district, and some small pockets in the eastern and northern part of the district. Moreover, the western and central part of the district is covered with the very low to low agricultural density villages. A simple linear relationship was drawn between the agricultural density villages and annual incidence rate of VL cases and showed a significant relationship (*rho* = 0.29, $p < 0.003$).

### 8.3.7 Association between Nutritional Density and Average Annual Incidence of VL

Nutritional density of Muzaffarpur district has ranged from 0.01 to 3.6 (SD±1.19). Based on the nutritional density values, the study area is divided into five classes using geometric interval. The area less than 0.050 nutritional values are considered as low nutritional density, while the values more than 0.30 is considered high nutritional density of villages. The maximum density records are of the southwestern and southern part of the district. Some small pockets of higher nutritional density were also recorded from the eastern and northern part of the study area (Figure 8.3f). The central part of the district has shown very low nutritional density. The spatial autocorrelation between the similar nutritional density villages showed clustered pattern (*Moran's* $I = 0.46$, Z-score 34.05, $p < 0.01$). The spatial relationship between the VL affected villages and nutritional density showed a significant relationship (*rho* $= 0.26$, $p < 0.033$).

### 8.3.8 Analysis of Kala-Azar Transmission Risk Model

Kala-azar transmission risk map of Muzaffarpur district derived through population characteristics is shown in Figure 8.4 and the relative importance of each criterion and their categorization is shown in Table 8.2. The risk index value has ranged from 12.5 to 48.0. Based on the geometric interval, the risk index value was categorized into (i) 'high' risk areas (risk index value >0.40), (ii) 'medium' risk areas (risk index values 34.55–39.39), (iii) 'low risk' areas (risk index values 25.12–34.54) and (iv) 'very low' risk areas (risk index value <25.12). The 'high risk' areas on the map are shown in 'dark black' colour, while the 'low risk' areas are shown in 'light grey'. However, the high risk areas were found mostly in the western, southern and some small pouch of eastern and northern part of the district. On the other hand, the central part of the area showed very low risk for kala-azar transmission.

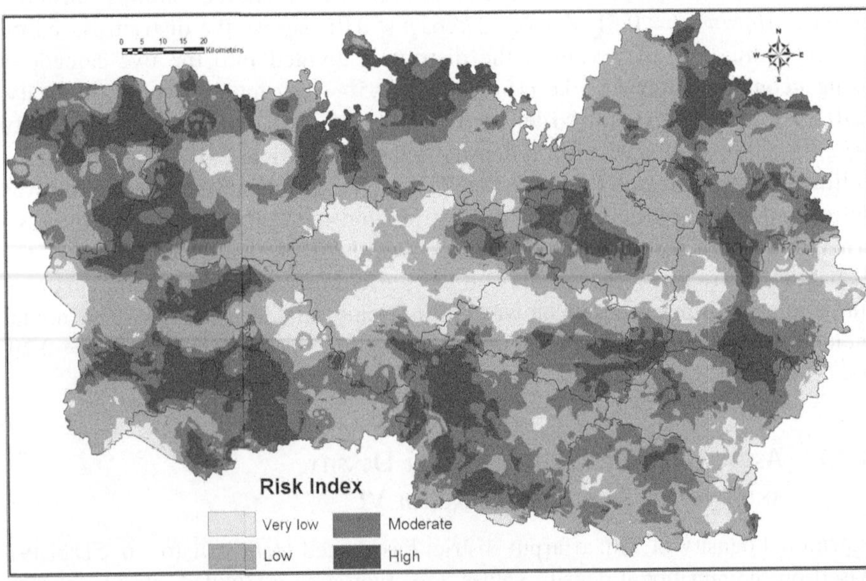


**FIGURE 8.4** Raster-based kala-azar transmission risk model developed through demographic characteristics of Muzaffarpur district

**TABLE 8.2**
**Raster-Based Index Model by Calculating Index Values and Summing Up Weighted Cell Values of Demographic Characteristics for Kala-Azar Transmission in Muzaffarpur District**

| Input Theme | Criteria of Weightage | Label | Weightage |
|---|---|---|---|
| Population density | 10 | Less than 100 | 1 |
| | | 101–150 | 2 |
| | | 151–200 | 3 |
| | | 201–250 | 4 |
| | | More than 250 | 5 |
| Family size | 30 | Less than 4.50 | 1 |
| | | 4.51–5.00 | 2 |
| | | 5.01–5.50 | 3 |
| | | 5.51–6.00 | 4 |
| | | More than 6.01 | 5 |
| Illiteracy rate | 25 | Less than 40.00 | 1 |
| | | 40.01–50.00 | 2 |
| | | 50.01–60.00 | 3 |
| | | 60.01–70.00 | 4 |
| | | More than 70.01 | 5 |
| Per cent of unemployed population | 15 | Very low | 5 |
| | | Low | 4 |
| | | Moderate | 3 |
| | | High | 2 |
| | | Very high | 1 |
| Agricultural density | 10 | Less than 0.50 | 1 |
| | | 0.51–1.00 | 2 |
| | | 1.01–5.00 | 3 |
| | | 5.01–10.00 | 4 |
| | | More than 10.00 | 5 |
| Nutritional density | 10 | Less than 0.050 | 1 |
| | | 0.051–0.100 | 2 |
| | | 0.101–0.150 | 3 |
| | | 0.151–0.300 | 4 |
| | | More than 0.301 | 5 |

## 8.4 DISCUSSION

The Muzaffarpur district of Bihar, India has veteran kala-azar epidemics during the twentieth century and assassination of thousands of poor people in each outbreak. To fight off the disease and put a stop to the spread of an epidemic, VL-infected individuals must undergo expensive medical treatment. Regrettably, the majority of the inhabitants of the study sites do not have access to affordable and accessible health care due to very poor socioeconomic conditions (Thakur, 2000; World Bank [WB], 1998).

Several hypotheses have been delineated by the qualitative and quantitative assessment that the VL disease may be strongly related to socioeconomic variables (Kesari et al., 2010; Pascual Martınez et al., 2012; Schenkel et al., 2006). In the present study, parameters like population density, family size, illiteracy rate, unemployed population, agricultural density and nutritional density were considered and their spatial relationship was delineated that may provide as a guideline on how to better allocate resources for various VL control measures and improving such factors can have the greatest impact on control of the spread of the disease. Alternatively, because of the patriarchal nature of Musahar communities in Bihar, literate persons can formulate better-informed health assessments for the family, and hence optimistically influence the prevalence of VL disease (Sheets et al., 2010).

Population density (i.e. population per square km) is positively correlated with the reported incidences during the period. The sand fly vector in India, *Phlebotomus argentipes*, is endophagic (e.g. biting occurs inside houses) and bites in the evening and night (WHO, 2001). On the other hand, expanding human population living in overcrowded conditions with inadequate housing and sanitary facilities increases the likelihood of VL infection (Arias et al., 1996; Costa et al., 1990; Mendes et al., 2002). Thus, for example, greater density in terms of household members per room could attract more sand flies, growing an individual family member's exposure to sand flies and, consequently, the possibility of acceptance of an infected bite (Reithinger et al., 2010).

Family size, in general, was also a significant variable for VL transmission in this region (Pascual Martınez et al., 2012; Schenkel et al., 2006). Our results showed that large family size positively associated with the average annual incidence of VL in the study site. This seems reasonable because when sand flies are more active, chances of sand fly bite and disease transmission increase. Earlier studies, conducted by Ranjan et al. (2005), reported that the incidences of VL were higher among cases with the past of kala-azar history among the family members in the past year. The presence of VL cases in the family might aid the transmission of this disease in the presence of sand fly vectors and other conditions favourable for completion of transmission cycle within the house (Bern et al., 2005).

The geographical difference of illiteracy rate (both the male and female) was considered as an important socioeconomic indicator for VL transmission. The population factors, such as literacy levels, economic eminences, on the occurrence of VL at the different scales have been depicted by other researchers and scientists (Adhikarı et al., 2010; Joshi et al., 2008; Ranjan et al., 2005). In earlier studies in South Asia, a strong association between kala-azar and poverty has been suggested (Bern et al., 2000; Thakur, 2000). The results of our analysis illustrated that the areas with unemployed population and higher illiteracy were measured highly threat areas for transmission of kala-azar. It may be due to the individuals who do not have acquaintance of the symptoms of the disease and are unconscious of disease prevention techniques (Hertzman, 1994; Ross and Wu, 1995). On the other hand, female literacy rate is crucial for understanding the impact of VL as females come forward for treatment only at the last stage of the disease, which is mainly due to the social and economic stigma (Singh et al., 2006).

Moreover, the higher unwaged population is considered as a proxy of income status of the geographical area that replicates a poorer living standard and escorts to an amplified risk for transmission. Badly constructed house are simply assaulted by sand flies and symbolize favourite sites for breeding (Schaefer et al., 1995). On an average, the household spends $134 on the treatment of family member suffering from kala-azar. This represents more than 50% of what is spent on a family member in an entire year (Thakur et al., 2008). The deficit of consequence of socioeconomic parameters may imitate the relative homogeneity of the study area and the fact that at the village level, more contiguous factors establish the risk for disease transmission.

Agricultural Density is imperative in geography primarily for economic reasons. It has proved as a useful index of man-land relationship primarily in an agrarian context (Quazi and Quazi, 2006). A higher agricultural density suggests that the available agricultural land (i.e. farms) is being used by more and may reach its output limit sooner than a nation that has a lower agricultural density. Conversely, an area with a low agricultural density essentially has a higher prospective for agricultural production. Our study illustrates most of VL cases occurred in the higher agricultural density region. Given that the significant relationship demonstrated between agricultural density and VL incidences should be considered when modelling the transmission risk areas for kala-azar.

Furthermore, it is a ratio between total population and cultivated area, expressed in terms of person per km$^2$ of cultivated land. Additionally, the nutritional density is thought to be an indication of living standards of any region. High nutritional density is may be due to the greater yield and multiple crops in a year. Our study showed, the value of nutritional density less than 0.15 is prone to high kala-azar transmission.

Finally, based on the proxy indicators of demographic variables, a composite raster-based index model was developed with a combination of all these factors (e.g. population density, family size, illiteracy rate, unemployed population, agricultural density and nutritional density), indicating the 'high' and 'low' risk of kala-azar transmission. Though, there are several limitations of this study, like (i) the variation in coverage of VL incidents over the years is probably leading to specious recognition of VL risk or non-risk areas; (ii) present study is based on secondary data available in a census record and as such the present analysis is based on passive case detection that may impede to rule out the opportunity of case underreporting. However, our model does not intend to assess the prevalence of disease, but only to recognize at the micro-level, and the areas showing a high risk potential for VL transmission in relation to population characteristics. Moreover, the model is based on the secondary database, and not any conventional field survey is required to collect the population characteristics. Similarly, remote sensing technique aids to estimate the crop and agricultural land of the study area, to delineate the agricultural and nutritional density. However, the risk map derived through the demographic proxy variables may help to control the kala-azar in delineating feasible areas of kala-azar transmission to strengthen the control strategy. It may also be suggested that all people alive with a risk of contagion need not be 'infective' due to other co-factors may be involved, such as the surrounding environment, reservoir and other biological factors. Nevertheless, a comparatively high risk of kala-azar transmission can be consistently expected with escalating values of 'risk' as delineated by this model. This information may be of value to health practitioners and health planners, who may employ this map as well as information to determine and plan for kala-azar supervision activities as well as to necessitate alleviation programmes for the focused intervention.

## ACKNOWLEDGEMENT

We are thankful to District Malaria Officer (DMO), Muzaffarpur, Bihar for his kind co-operation and freely providing the disease incidence report. We are also grateful to Dr. Nandini Chatterjee, Dr. Shreekant Kesari for their guidance and support. We are also grateful to the Census of India, Patna, Bihar for their kind assistance. We would like to thank USGS Earth Explorer community for freely providing satellite data.

## REFERENCES

Actor P 1960. Protein and vitamin intake and visceral leishmaniasis in the mouse. *Experimental Parasitology* 10: 1–20.

Adhikari SR, Maskay NM, Sharma BP 2009. Paying for hospital-based care of kala-azar in Nepal: Assessing catastrophic, impoverishment and economic consequences. *Health Policy and Planning* 24: 129–139.

Adhikari SR, Supakankunti S, Khan MM 2010. Kala-azar in Nepal: Estimating the effects of socioeconomic factors on disease incidence. *Kathmandu University Medical Journal* 8: 73–79.

Alvar J, Yactayo S, Bern C 2006. Leishmaniasis and poverty. *Trends in Parasitology* 22: 552–557.

Arias JR, Monteiro PS, Zicker F 1996. The reemergence of visceral leishmaniasis in Brazil. *Emerging Infectious Diseases* 2: 145–146.

Assimina Z, Charilaos K, Fotoula B 2008. Leishmaniasis: An overlooked public health concern. *Health Science Journal* 2(4): 196–205.

Banai-Kashani R 1989. A new method for site suitability analysis: The analytic hierarchy process. *Environmental Management* 13: 685–693.

Bern C, Hightower AW, Chowdhury R, Ali M, Amann J, Wagatsuma Y, Haque R, Kurkjian K, Vaz LE, Begum M, Akter T, Cetre-Sossah CB, Ahluwalia IB, Dotson E, Evan Secor W, Breiman RF, Maguire JH 2005. Risk factors for Kala-azar in Bangladesh. *Emerging Infectious Disease* 11(5): 655–662.

Bern C, Joshi AB, Jha SN, Das ML, Hightower A, Thakur GD, Bista MB 2000. Factors associated with visceral leishmaniasis in Nepal: Bed-net use is strongly protective. *American Journal of Tropical Medicine and Hygiene* 63: 184–188.

Boelaert M, Meheus F, Sanchez A, Singh SP, Vanlerberghe V, Picado A, Meessen B, Sundar S 2009. The poorest of the poor: A poverty appraisal of households affected by visceral leishmaniasis in Bihar, India. *Tropical Medicine and International Health* 14: 639–644.

Câmara G, Carvalho MS, Cruz OG, Correia V. Análise espacial de áreas. In: Druck S, Carvalho MS, Câmara G, Monteiro AMV, editores. Análise espacial de dados geográficos [monografia on-line]. São Paulo: Instituto Nacional de Pesquisas Espaciais; 2002. p. 1–32

Census of India, 2001. Statistics of Bihar. Census of India.

Cerf BJ, Jones T. C, Badaro R, Sampaio D, Teixeira R, Johnson WD Jr 1987. Malnutrition as a risk factor for severe visceral leishmaniasis. *Journal of Infectious Diseases* 156: 1030–1033.

Costa CH, Werneck GL, Rodriques L Jr et al. 2005. Household structure and urban services: Neglected targets in the control of visceral leishmaniasis. *Annals of Tropical Medicine and Parasitology* 99: 229–236.

Costa CHN, Pereira HF, Araújo MV 1990. Epidemia de leishmaniose visceral no Estado do Piauí, Brasil, 1980–1986. *Rev Saúde Pública* 24: 361–372.

de Almeida AS, de Andrade Medronho R, Werneck GL 2011. Identification of risk areas for Visceral Leishmaniasis in Teresina, Piaui State, Brazil. *American Journal of Tropical Medicine and Hygiene* 84(5): 681–687.

Desjeux P 2001. The increase in risk factors for leishmaniasis worldwide. *Transaction of the Royal Society of Tropical Medicine and Hygiene* 95: 239–243.

Dye C, Williams BG 1993. Malnutrition, age and the risk of parasitic disease: Visceral leishmaniasis revisited. Proceedings of the Royal Society of London, Series B 254: 33–39.

Fisher NI, Lewis T, Embleton BJJ 1987. Statistical analysis of spherical data. Cambridge: Cambridge University Press.

Herbreteau V, Salem G, Souris M, Hugot J-P, Gonzalez J-P 2005. Sizing up human health through remote sensing: Uses and misuses. *Parassitologia* 47: 65–81.

Hertzman C 1994. The lifelong impact of childhood experience: A population health perspective. *Dædalus* 123(4): 167–180.

Hu W, Tong S, Mengersen K, Oldenburg B 2007. Exploratory spatial analysis of social and environmental factors associated with the incidence of Ross River virus in Brisbane, Australia. *American Journal of Tropical Medicine and Hygiene* 76: 814–819.

Joshi A, Narain JP, Prasittisuk C, Bhatia R, Hashim G, Jorge A, Banjara M, Kroeger A 2008. Can visceral leishmaniasis be eliminated from Asia? *Journal of Vector Borne Diseases* 45: 105–111.

Kesari S, Bhunia GS, Kumar V, Jeyaram A, Ranjan A, Das P 2010. Study of house-level risk factors associated in the transmission of Indian Kala-azar. *Parasit & Vectors* 12(3): 94.

Loh JM 2011. K-scan for anomaly detection in disease surveillance. *Environmetrics* 22: 179–191.

Malczewski J 2000. On the use of weighted linear combination method in GIS: Common and best practice approaches. *Transaction in GIS* 4: 5–22.

Meheus F, Boelaert M, Baltussen R, Sundar S 2006. Costs of patient management of visceral leishmaniasis in Muzaffarpur, Bihar, India. *Tropical Medicine and International Health* 11: 1715–1724.

Mendes WS, Silva AA, Trovão JR, Silva AR, Costa JM 2002. Expansão espacial da leishmaniose visceral americana em São Luis, Maranhão, Brasil. *Revista da Sociedade Brasileira de Medicina Tropical* 35: 227–231.

Pascual Martınez F, Picado A, Roddy P, Palma P 2012. Low castes have poor access to visceral leishmaniasis treatment in Bihar, India. *Tropical Medicine and International Health* 17(5): 666–673.

Quazi SA, Quazi NS 2006. Population Geography. APH Publishing Corporation, New Delhi, ISBN: 81-7648-993-X.

Ranjan A, Sur D, Singh VP, Siddique NA, Manna B, Lal CS, Sinha PK, Kishore K, Bhattacharya SK 2005. Risk factors for Indian kala-azar. *American Journal of Tropical Medicine and Hygiene* 73(1): 74–78.

Reithinger R, Mohsen M, Leslie T 2010. Risk factors for anthroponotic cutaneous leishmaniasis at the household level in Kabul, Afghanistan. *PLoS Neglected Tropical Diseases* 4(3): e639. doi:10.1371/journal.pntd.0000639

Rijal S, Koirala S, van der Stuyft P, Boelaert M 2006. The economic burden of visceral leishmaniasis for households in Nepal. *Transactions of the Royal Society of Tropical Medicine and Hygiene* 100: 838–841.

Roland A 2002. Infectious diseases on the 21st century. *Infectious Circular – WHO Mediterranean Zoonoses Control Centre* 55: 1–16.

Ross CE, Wu C 1995. The links between education and health. *American Sociological Review* 60: 719–745.

Sarnoff R, Desai J, Desjeux P, Mittal A, Topno R, Siddiqui NA, Pandey A, Sur D, Das P 2010. The economic impact of visceral leishmaniasis on rural households in one endemic district of Bihar, India. *Tropical Medicine and International Health* 15(Suppl. 2): 42–49.

Satty TL 1980. The Analytic Hierarchy Process. McGraw-Hill, New York.

Schaefer KU, Kurtzhalz JA, Gachihi GS, Muller AS, Kager PA 1995. A prospective sero-epidemiological study of visceral leishmaniasis in Baringo district, Rift Valley Province, Kenya. *Transactions of the Royal Society of Tropical Medicine and Hygiene* 89: 471–475.

Schenkel K, Rijal S, Koirala S, Koirala S, Vanlerberghe V, Van der Stuyft P, Gramiccia M, Boelaert M 2006. Visceral Leishmaniasis in southeastern Nepal: A cross-sectional survey on *Leishmania donovani* infection and its risk factors. *Tropical Medicine & International Health* 11(12): 1792–1799.

Sharma A, Bern C, Varghese B, Chowdhury R, Haque R, Ali M, Amann J, Ahluwalia IB, Wagatsuma Y, Breiman RF, Maguire JH, McFarland DA 2006. The economic impact of visceral leishmaniasis on households in Bangladesh. *Tropical Medicine & International Health* 11: 757–764.

Sharma BP, Maskay NM, Adhikari SR, Andrews JR, Joshi AB, Wijeyaratne P, Joshi SD 2004. Socio–economic determinants of kala-azar in Nepal. *Journal of the Nepal Health Research Council* 2: 35–42.

Sheets D, Mubayi A, Kojouharov HV 2010. Impact of socioeconomic conditions on the incidence of visceral leishmaniasis in Bihar, India. *International Journal of Environmental Health Research* 20(6): 415–430.

Singh J, Dhillon SS 2004. Agricultural Geography. 3rd Edition. Tata McGraw-Hill Publishing Company Limited. ISBN: 0-07-053228-1.

Singh SP, Reddy DC, Rai M, Sundar S 2006. Serious underreporting of visceral Leishmaniasis through passive case reporting in Bihar, India. *Tropical Medicine & International Health* 11(6): 899–905.

Sundar S, Arora R, Singh SP, Boelaert M, Varghese B 2010. Household cost-of-illness of visceral leishmaniasis in Bihar, India. *Tropical Medicine and International Health* 15(Suppl. 2): 50–54.

Thakur CP 2000. Socio economics of visceral leishmaniasis in Bihar (India). *Transactions of the Royal Society of Tropical Medicine and Hygiene* 94(2).156–157.

Thakur CP, Thakur S, Narayan S, Sinha A 2008. Comparison of treatment regimens of kala-azar based on culture & sensitivity of amastigotes to sodium antimony gluconate. *Indian Journal of Medical Research* 127: 582–588.

Trewartha GT 1953. A case for population geography. *Annals of the Association of American Geographers* 43(2): 71–97.

Willmott CJ 1984. On the evaluation of model performance in physical geography. In Spatial Statistics and Models, ed. G. L. Gaile, and C. J. Willmott, pp. 443–460. Springer, Dordrecht Netherlands.

World Bank (WB) 1998. Reducing Poverty in India: Options for More Effective Public Services. World Bank, Washington, DC.

World Health Organization (WHO) 2001. WHO Recommended Strategies for the Prevention and Control of Communicable Diseases. Department of Communicable Disease Control, Prevention and Eradication. WHO/CDS/CPE/SMT/2001.13. Available at: http://whqlibdoc.who.int/hq/2001/WHO_CDS_CPE_SMT_2001.13.pdf

World Health Organization (WHO) 2002. Urbanization: An increasing risk factor for leishmaniasis. *Weekly Epidemiological Records* 77: 365–372.

Zarina LM, Lebedev SV, Nesterov EM 2010. Ecological geochemical investigations of the contents of heavy metals in the snow cover in the Saint-Petersburg region with application of GIS technologies. *2010 International Conference on Biology, Environment and Chemistry (IPCBEE), vol. 1 (2011), IACSIT press, Singapore*, pp. 40–43.

# 9 Hydropower Site Optimization for Sustainable Energy Supply

## *A Case Study in PNG*

*Tingneyuc Sekac, Sujoy Kumar Jana, and Nosare Maika*

## 9.1 INTRODUCTION

Most countries that have full supply (90–100%) of electricity are experiencing dramatic increase in their economic sector and prosperity. In order to improve people's life in rural areas, the reliable and affordable electricity access is a basic requirement. The healthcare and education systems processes including economic growth will be greatly improved through available and reliable electricity. Electrification development and roll out needs close consideration in terms of investment opportunity and sustainability. Sustainable electricity or energy involves the responsible generation, distribution, and utilization of power resources to meet current needs while safeguarding the ability of future generations to meet their own needs. The sustainable electricity production revolves around harnessing renewable energy sources like solar, wind, hydro, and geothermal power, which are naturally replenished and emit minimal greenhouse gases. The practices prioritize energy efficiency, conservation, and the reduction of environmental impact. This includes promoting the adoption of clean technologies, improving energy infrastructure, and implementing policies that support renewable energy deployment. Additionally, sustainable energy initiatives aim to foster social equity, economic development, and environmental stewardship, ensuring that energy access is affordable, reliable, and equitable for all communities. By transitioning toward sustainable energy systems, societies can reduce reliance on finite fossil fuel resources, mitigate climate change, and build a more resilient and prosperous future.

Thus according to Nerini et al. (2016), thorough planning for electrification options, site location, construction, and distributions is required. The planning should aim at being sustainable from social point of view, economic and environment point of view (Neves and Leal, 2010). Electricity is a vital resource for populations, businesses, and industries worldwide. Presently, emerging and advanced technologies offer alternatives to grid-connected power, by exploring other renewable energy sources. Adoption

DOI: 10.1201/9781032683928-9

of renewable energy offers several substantial advantages, including a reduction in the consumption of fossil fuels and the harnessing of readily available resources. There are concerted efforts to extend electricity access to even the most remote areas, with a focus on leveraging hybrid connection systems that integrate renewable energy sources while still incorporating the conventional grid infrastructure. Furthermore, Papua New Guinea (PNG) is among the countries classified as least developed in terms of infrastructure and the provision of essential services and basic necessities.

It has been emphasized that the challenges associated with electrification production and supply to remote areas in PNG are multifaceted. These constraints are primarily attributed to the extensive distances required for electricity transmission lines and pole distribution, the intricate and challenging topography that complicates the installation of such infrastructure, resulting in elevated costs. Additionally, issues related to land ownership, the region's low population density, limited education levels, low load density, and insufficient revenues further compound the challenges (Kaur and Segal, 2016). Furthermore, the absence of robust market infrastructure hinders the attraction of private investors for power production and supply. PNG government has now established national electrification policy (NEP) to have control over on and create pathways and solutions toward electrification roll out to rural places (PNG National Energy Policy [NEP], (2017). This is in line to supporting and enforcing national pillars put forward and vision 2050 to supply power from renewable energy and sustainable energy sources (NEP, 2017). The more possible means of establishing rural electrification can be through hybrid system development as illustrated in Figure 9.1. The other possible means of rural electrification supply can be

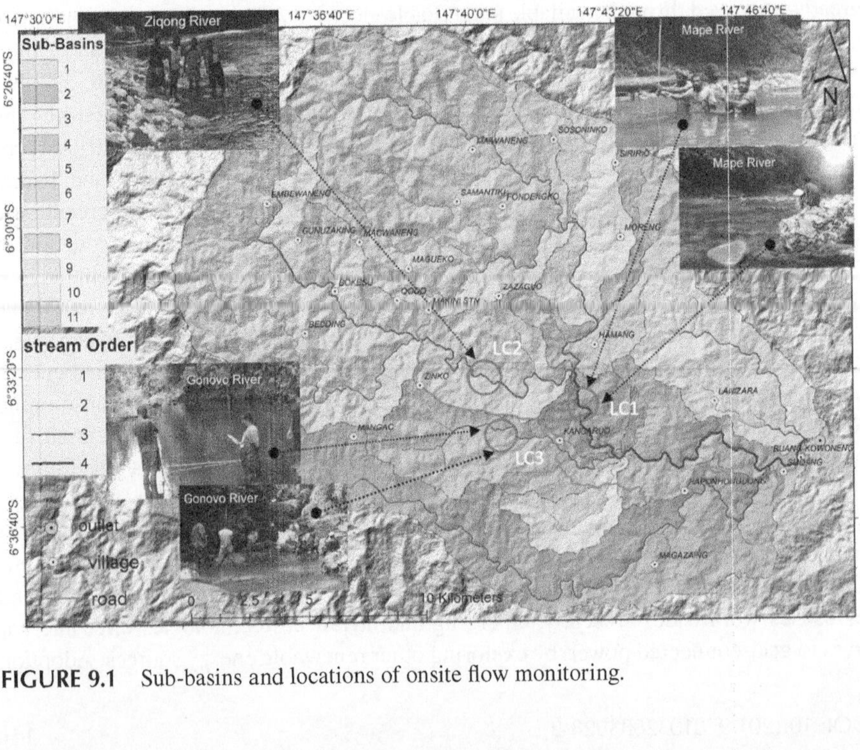


FIGURE 9.1 Sub-basins and locations of onsite flow monitoring.

through national government capital cost subsidies of funding to reduce installation cost of the generators and grid distribution, therefore, the affordable tariff can be charge especially for rural communities, so they can afford to pay to maintain the supply and distribution.

In order to develop and establish rural electrification program within such complex environment as PNG, taking research approaches is the key platform. Through such approach, meaningful pathways can be establishing to achieve the goals of 70% electrification by 2030 and 100% by 2050. More data and information can be populated via research output to aid efficient planning and decision-making for rural electrification. Through research and development, potential of renewable energy can be estimated including feasible sites for construction and supply, evaluation of efficacy of generators, hydro turbine, solar panels, battery storage, and evaluation of available natural energy. In line to that, the most common important factor is the population and/or socio-economic statues of general communities. Through research the population behavior, responses, population distribution, rate of household earnings, etc., are to be fully evaluated to be understood and at the same time it is as a means of awareness that is addressed to general communities.

The current research represents a significant step, among various initiatives, in the pursuit of viable strategies for rural electrification. The research primarily focuses on an in-depth assessment of renewable energy potential. Taking Mape catchment in Finschhafen District, PNG as a case study zone, the study extensively evaluates economic status and conditions of the communities or villages residing in this catchment area. Additionally, the research entails a thorough investigation into the calculation and estimation of household power consumption rates and electricity demand within the communities. Moreover, the objective of this research looks into evaluating the potential of hydropower and establishes correlations with the electricity demand, affordability, and sustainability of electricity provision for these communities. To estimate the hydropower potential, the study predominantly involves preliminary calculation of flow discharge through utilizing soil and water assessment tool (SWAT) Cüceloğlu et al. (2021), and Strauch et al. (2012), for the catchment region and conducting a detailed topographical analysis. In the case of electrification consumption rate and demand, as well as the assessment of the communities' economic status, household survey data serves as the primary input for computation.

The case study was carried out in the Mape catchment region in the Finschhafen District of PNG. This expansive region covers 432.10 square kilometers and, as per data from the 2011 census and recent field surveys, is home to an estimated population of approximately 7,621 plus residents. The study area primarily consists of mountainous terrain with an intricate network of streams and rivers. Accessibility is a significant challenge, with most areas lacking proper road connections. Transportation, in many parts of the region, primarily relies on foot travel. Communities within the study area do not have access to a centralized electricity supply. Instead, they rely on privately owned solar panels and personal generators for lighting and charging. The predominant cash crops cultivated in this region are coffee, cocoa, and vanilla. Rice cultivation also plays a dual role, serving as both a source of income and a staple food for daily consumption. Approximately 80% of the study region is covered by forested land, while 60% is dedicated to agricultural or crop cultivation, indicating a deliberate commitment to gardening practices. Some areas feature low-density vegetation. Approximately 2% of the study region comprises built-up or barren land, serving as zones for village communities, road networks, and other infrastructure.

## 9.2 RESEARCH METHOD

To estimate power potential at specific locations, various data layers were collected, processed, and analyzed. The key input data for estimating power potential included flow discharge in cubic meters per second ($m^3/s$) and effective head in meters (m). Flow discharge within the catchment region was estimated using the QSWAT plugin in QGIS 3.16.1 and was calibrated and validated in the SWAT-CUP platform. Researchers incorporated three years of on-ground field observation flow data for calibration and validation. The research further focused on evaluating the performance of the SWAT model in flow simulation. Model performance evaluation was conducted at three locations within the study area catchment region – Location-1 (LC1), LC2, and CL3 (refer to Figure 9.1). The Nash-Sutcliffe Efficiency (NSE) and correlation coefficient ($R^2$) methods were used for this evaluation. The below equation highlights the processes of model performance evaluation

$$NSE = \frac{\Sigma_{i=0}^{n}\left(Q_m - Q_s\right)^2}{\Sigma_{i=0}^{n}\left(Q_m - \bar{Q}_m\right)^2} \quad R^2 = \frac{\left[\Sigma_{i=0}^{n}\left(Q_m - \bar{Q}_m\right)\left(Q_s - \bar{Q}_s\right)\right]^2}{\Sigma_{i=0}^{n}\left(Q_m - \bar{Q}_m\right)^2 \Sigma_{i=0}^{n}\left(Q_s - \bar{Q}_s\right)^2},$$

where:

$Q_m$ = Measured/Observed flow (cms).
$Q_s$ = Simulated Discharge (cms).
$\bar{Q}_m$ = Mean Measured/Observed flow (cms).
$\bar{Q}_s$ = Mean Simulated Discharge (cms).

Data inputs for flow estimation within QSWAT primarily comprised climate data (Sekac et al., 2021), topographic data, and soil data, including Land Use Land Cover (LULC). Data for effective head estimation (in meters) was obtained from the Digital Elevation Model (DEM) provided by the Shuttle Radar Topography Mission (SRTM). Additional data types utilized in the research included field data, such as household GPS locations, household socio-economic data, household electricity load and demand, and information on electrical appliances. This data was used for evaluating economic potential and assessing electricity supply-demand.

Estimating hydropower potential involves determining discharge in cubic meters per second and effective head in meters. Discharge estimation was accomplished using the SWAT model, while the DEM was employed for effective head (He) estimation. Within the catchment region, the stream network and topography were assessed to identify the inlet point, assumed to be the dam or weir, and the outlet point, assumed to be the power house location with its tailrace. The difference in elevation between these points was considered the effective head (He) for power calculations, using the equation: Power (kW) = p.g.Q.He.η, where 'Q' is the discharge in $m^3/s$, 'p' and 'g' are water density and gravity, He is the effective head, and η is turbine efficiency.

Additionally, turbine efficiency was considered based on the turbine efficiency curve under varying conditions of total discharge (Q) and effective head (He) (Carravetta et al., 2013). The type of turbine installed at the outlet or powerhouse section significantly influences turbine efficiency. Various turbine types are used in mini, small, and large hydropower systems, and selecting the most appropriate turbine is crucial during hydropower development design and implementation processes (Sammartano et al., 2019). Turgo and Pelton turbines are commonly used for high heads exceeding 50 m with small flow rates, while Francis and Crossflow turbines are more suitable for heads lower than 50 m with a wide range of flow rates. For heads less than 10 m with a high-flow rate, the Kaplan turbine is employed (Paish, 2002).

In identifying viable sites for hydropower constructions, considerations included hydrological constraints, topographical variations, environmental factors, economic considerations, and proximity to infrastructure (Kouadio et al., 2022). Field observations played a pivotal role, with the research team extensively exploring various locations within the Mape catchment region (refer to Figure 9.1). The objective was to gather data and delineate potential sites suitable for constructing small to large hydro dams. Tools such as GPS, questionnaire forms, digital flow meters, and UAV drones facilitated efficient data collection. Figure 9.1 visually showcases the Mape catchment region in the Finschhafen District of PNG and potential locations where flow is continuously monitored.

Building upon field observations and guided by specified criteria, the research team meticulously selected feasible potential sites at each location. The site selection process progressed from upstream to downstream, adhering to criteria such as maintaining a good distance between inlet and outlet (200–300 m for large hydro dams), ensuring an effective head of 20 m, a threshold flow accumulation value of 3,000 and above, and the feasibility of major stream networks. Additionally, nearby topography, slope, and proximity to communities and infrastructure (e.g., roads) were crucial considerations.

Another primary goal of the present research was to assess the load or demand, including the socio-economic conditions of the rural communities in the study region. According to Singh (2009), the most accurate method for assessing household demand for rural electrification involves direct community interaction. Singh (2009) emphasized using the Participatory Rural Appraisal (PRA) method for such community interactions. This approach enhances the efficiency and flexibility of electrification development and supply, fostering sustainability, reliability, viability, security, and agreement on willingness to pay.

During the demand assessment, the survey team actively engaged with communities through field visits to at least six villages, employing diverse methods to collect a range of data types, as illustrated in Figures 9.1 and 9.2. Utilizing questionnaire forms and interview techniques within the PRA method, the team gathered extensive household data, including population, monthly earnings, cash crops cultivated,

(a) Community discussions for Improving cash crop growing and marketing

(b) Orthophoto and DEM Capture and generation—UAV flight

(c) Hydropower Potential sites investigation

(d) Household Survey

(e) Peak Flood Monitoring

**FIGURE 9.2** Common task executed during field visit within Mape catchment region; (a) community gathering and discussion, (b) UAV data capture, (c) hydropower potential sites investigation, (d) household survey, and (e) peak flood monitoring.

income sources, future SME plans, anticipated electrical appliances, and willingness to pay for electricity (at a rate of 0.50 t–0.73 t kWh). Additionally, the survey team geotagged each household's location with GPS points.

Researchers conducted community gatherings and discussions in each village to proactively raise awareness about the potential benefits, sustainability, and tariff plans associated with electricity supply and demand. Complementing these efforts, a UAV drone was used to capture aerial footage of all the visited villages. The collected data underwent rigorous refinement and assessment for effective presentation and visualization. The insights gleaned through the PRA method are integral to planning and decision-making across various facets of rural electrification, emphasizing the community-centric approach adopted during the survey.

## 9.3 RESULTS AND DISCUSSIONS

A larger catchment region increases the likelihood of having sufficient water availability for successful and sustainable hydropower development. During the field survey, every generated stream or river network were examined. Refer to Figure 9.1, where observations based on field insights and GIS software calculations using flow accumulation reveal that order 1 streams are classified as small streams, while orders 2, 3, and 4 are classified as medium to major continuous flowing rivers.

The results of the calibration and validation were found to be satisfactory. Consequently, the discharge data was incorporated into power calculations, first in kilowatts (kW) and then converted to megawatts (MW) for presentation purposes.

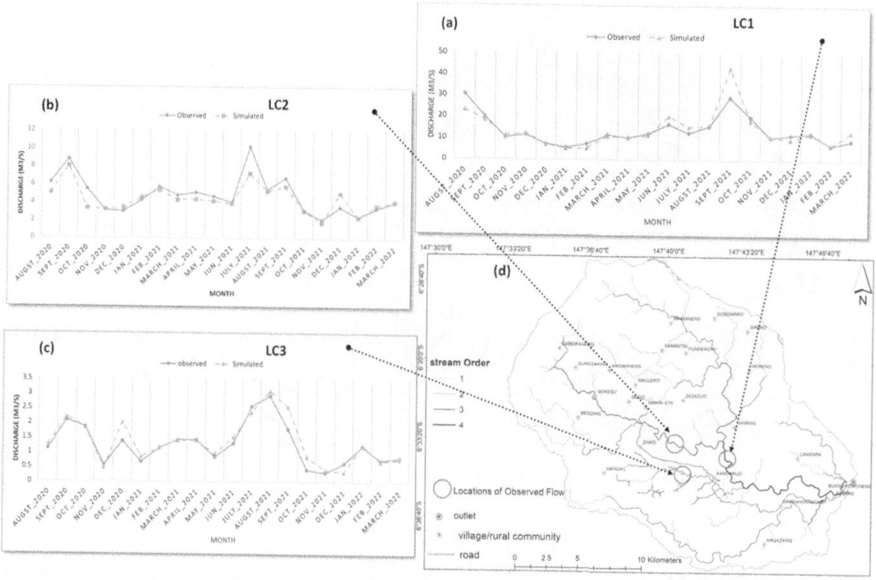


**FIGURE 9.3** Calibration and validation results; (a) Location 1 (LC 1), (b) Location 2 (LC2), (c) Location 3 (LC3), and (d) observed flow location map.

The calculated, calibrated, and validated discharge used for power calculations is shown in Figure 9.3. Figure 9.4 illustrates the validation and calibration results at each specific location where the flow was observed.

The NSE calibration statistical analysis results for observed and simulated flow are 0.75, 0.87, and 0.77 for LC1, LC2, and LC3, respectively. Additionally, the NSE validation statistical analysis results for observed and simulated flow are 0.89, 0.78, and 0.87 for LC1, LC2, and LC3, respectively. In the case of $R^2$ calibration, the statistical analysis results for observed and simulated flow are 0.94, 0.79, and 0.85 for LC1, LC2, and LC3. Furthermore, the $R^2$ validation statistical analysis results for observed and simulated flow are 0.88, 0.83, and 0.89 for LC1, LC2, and LC3. Mostly near-perfect linear correlations are observed including satisfactory NSE results. LC3 shows major deviation between observed and simulated within the month of August and November 2021 and including month of December 2020. L2 shows major deviation from August to November 2020 and July 2021. These deviations, observed at various locations, were a result of limited observed data integrated with a 22-year climate data range data processing and analysis. Continuous recording of observed data, integrated into the model, can improve accuracy and enhance agreement between observed and simulated flow.

To identify potential sites and specify the head (He) within the catchment region, criteria were employed as a guide, and the potential head in meters (m) was calculated. Each stream/river network were individually assessed from its origin to the endpoint. Refer Figures 9.1 and 9.2 for the stream or river network. A specific site within the stream/river network is considered a potential site when it fulfills the

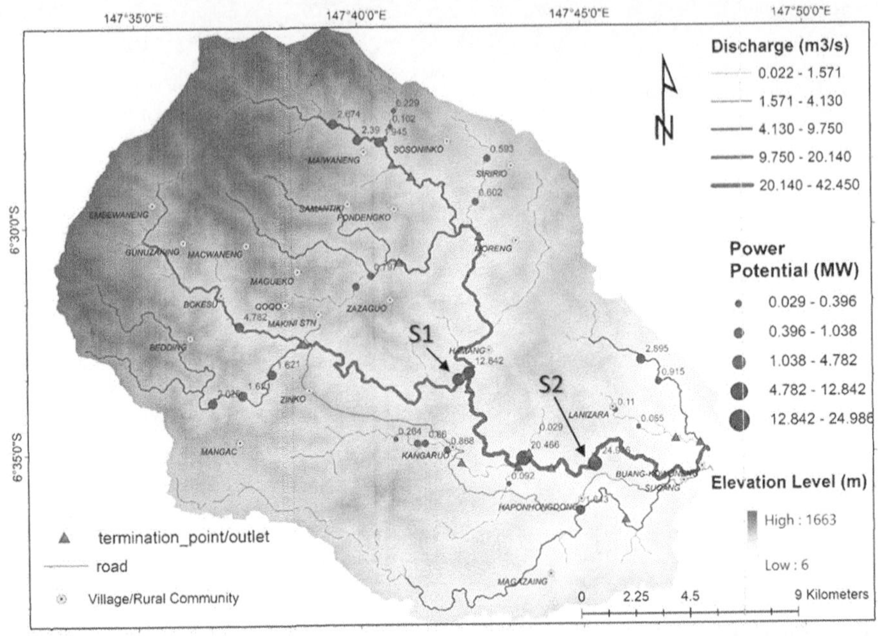


**FIGURE 9.4** Estimated discharge and calculated power potential in MWh.

criteria. The initial upstream site is designated as the weir, and the subsequent downstream site is recognized as the outlet or powerhouse. As one moves downstream, the site is deemed a termination point when stream or river end point is reach. Refer Figure 9.4.

In the preliminary stage, a topography assessment for planning the layout of run-of-river hydropower infrastructure was conducted using various criteria. Sites that aligned with the topographical settings and other specified criteria were deemed viable. For these viable sites, discharge values were extracted, and the effective head was calculated. Subsequently, the power potential of these sites was calculated in megawatts (MW), as shown in Figure 9.4. The calculated power ranges from 0.0291 to 24.9862 MW, with smaller streams exhibiting lower power output and larger rivers showing higher power output within the study region. The primary factors influencing power output in MW were flow discharge (Q) and effective head (He). Sites closer to villages are considered more valuable, as the communities can benefit from the generated power. This consideration led to another objective of the study: estimating the power consumption rate and demand of villages or rural communities within the catchment region.

In the course of assessing and calculating for the run-of-the-river hydropower scheme, two potential or viable sites emerged as options for a large hydropower dam, The sites are denoted as S1 and S2, refer Figure 9.4. This choice was influenced by the notably steep slope between these closely situated sites. The power output is from 8 to 23 MW. Majority of other sites are being categorized as run-of-the-river hydropower schemes.

### 9.3.1 Socio-Economic and Electricity Supply-Demand

A primary objective of the current research was to evaluate the viability and affordability of utilizing a hydropower system to meet the electricity demand of local communities. The study aimed to determine the power needs of nearby communities, with four communities selected as representative samples for detailed analysis. Figure 9.5 illustrates the Kangaruo community, which comprises 84 households along with public infrastructure. The other communities were examined using the same methodologies. The study involved evaluating key factors such as household earnings, population size, electric appliances and load, income sources, and demand. Furthermore, the study sought to assess each household's interest in connecting electricity to their homes and to evaluate whether households possess sufficient resources for sustaining and supporting long-term electricity supply. A total of 379 households within four rural communities, including their existing public infrastructures were

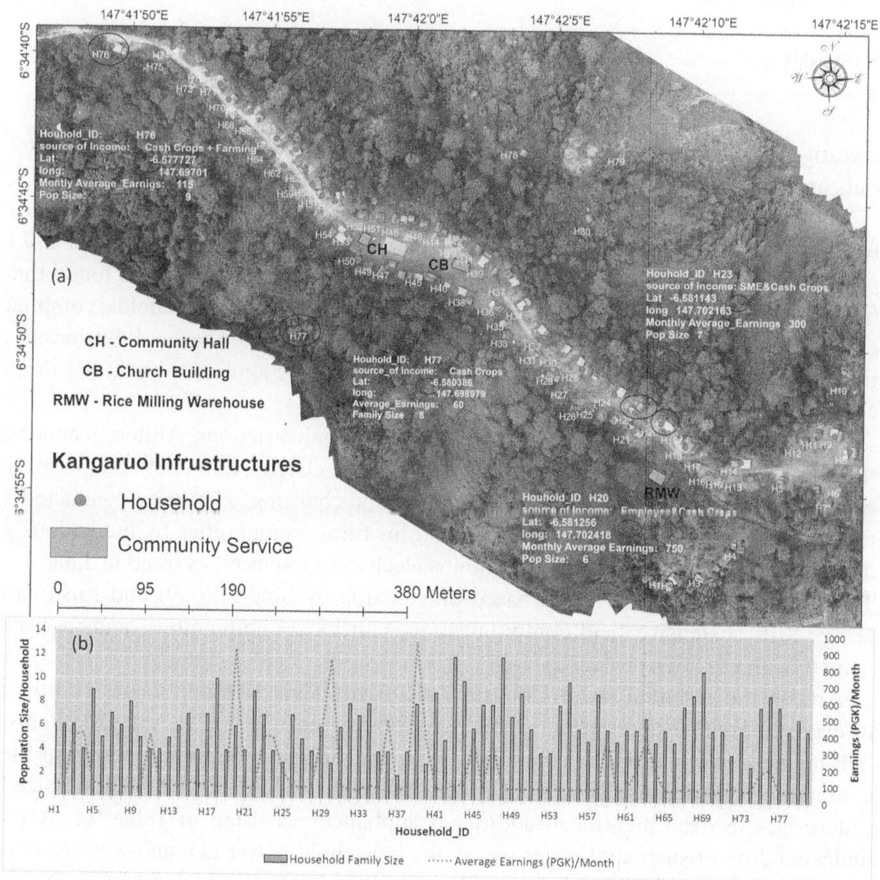


**FIGURE 9.5** Kangaruo community household data: (a) household location and (b) population size and earnings per month.

**TABLE 9.1**
**Electrical Appliances Average Ratings for Kangaruo Community**

| Electrical Appliances | No. House/ Institute | Appliances per House/Institute | Appliances Rate (Wattage) on Average |
|---|---|---|---|
| Lighting | 84 | 4 | 35 |
| DVD Player + TV (24″) | 50 | 1 | 53 |
| Laptop | 30 | 1 | 65 |
| Electric Kettle (2.5 L) | 20 | 1 | 1,200 |
| Electric sewing machine | 18 | 1 | 100 |
| Mobile phone charging | 82 | 2 | 25 |
| Sound system | 65 | 1 | 100 |
| Refrigerator | 15 | 1 | 300 |
| Printer | 2 | 1 | 200 |
| LED sport light | 4 | 3 | 100 |
| Workshop power tools | 15 | 1 | 1,500 |
| Water pump | 1 | 1 | 1,000 |
| Street lights | 1 | 30 | 20 |

investigated. Most housing type in the community is a mix of traditional bush materials and modern building materials with metal roofs.

For the Kangaruo community, Figure 9.5a and b illustrates the spatial distribution of household, average household size and monthly earnings, while Table 9.1 provides details on the quantity of appliances used per household. It was found that 62 households generate income through cash crop sales, 7 households combine employment with cash crop cultivation, and 10 households operate small businesses, including crop cultivation. The same methodologies were applied to the other three communities.

The demand assessment did not account for seasonal variations. Although none of the households in the study region have access to electricity, a few rely on privately owned generators and solar panels for lighting and charging. The study conducted a thorough evaluation of energy load demand for rural communities by meticulously developing and assessing a range of common electrical appliances, as listed in Table 9.1. These appliances were selected based on research by Singh (2009) and Rushman et al. (2019), extensive field communication with local communities, and insights gained from previous projects.

Among the essential and widely used electrical devices or appliances common to every household and certain public institutions are televisions, sound systems, lighting systems, and phone chargers. These appliances were chosen based on the community's responses. Respondents who had plans for Small and Medium-sized Enterprises (SMEs) mentioned additional appliances as listed in Table 9.1. After understanding of such appliances usage, the household power demand was calculation. The calculated power demand for four villages are illustrated in Figure 9.6. Hopohondong communities with 232 household, its total power demand stands at 311 kW. 121, 73, and 68 kW for Kangaruo, Manga, and Zinko community. Figure 9.4

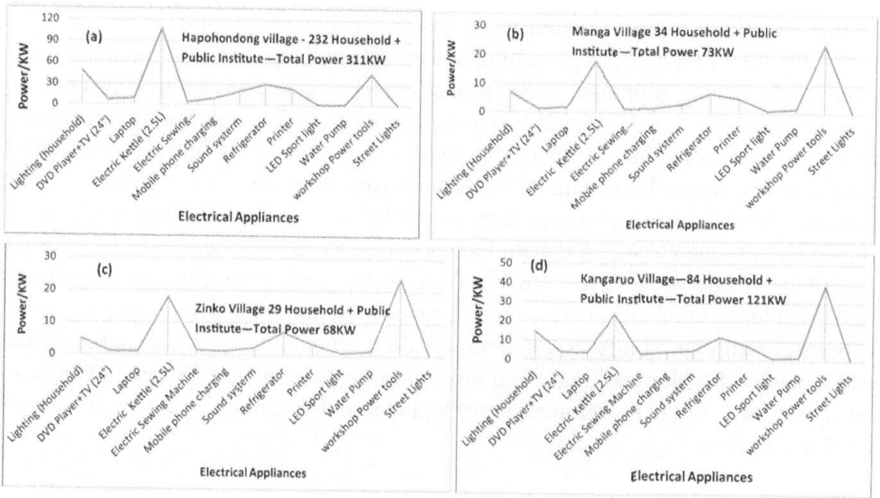


**FIGURE 9.6** Rural communities power demand: (a) Hapohondong community, (b) Manga community, (c) Zinko community, and (d) Kangaruo community.

illustrates the calculated power potential locations and also illustrates the proximity of each communities to potential sites.

## 9.4 CONCLUSION AND RECOMMENDATION

The current research aims to create pathways to sustainable rural electrification development with clean energy solution. Most of the natural flowing streams and rivers are left unexplored for hydropower potential. Further to that, other energy resources like solar, wind, etc. ... are also left unexplored at various locations. Carrying out research studies to evaluate and understand such energy resources and their potentiality will have greater positive impact toward near future for electrification development throughout country PNG. Climate change impact issues are on rise in the modern era. Having to create solutions to reduce such impact is on high demand. The ongoing study encompasses two phases. The primary phase involves the calculation and identification of potential hydropower sites, followed by the second phase, which focuses on evaluating household energy demand. Both desktop survey and field survey techniques were conducted to execute the task to its completion. Through a meticulous analysis process, 29 of the most viable potential sites were thoughtfully selected, and power potential calculated. The calculated power exhibits a range from 29 kW to an impressive 25,000 kW. Subsequent to the identification of these hydropower potential sites, the calculations seamlessly extended their scope to assess the intricate energy load demand of a communities within the study region. Total of 379 households were assessed to have total power demand at 573 kW. Same approaches were adopted for other communities within the study region for their power need. The proposed implementation of tariff rules serves as a strategic mechanism to proactively manage power usage. Furthermore, additional control measures are anticipated to be instituted, including appliances control or establishment of regulations governing household appliance usage to facilitate a balanced load-sharing approach. These proactive steps are envisioned to play a pivotal role in alleviating stress related to payment concerns. The research methodology employed a comprehensive strategy to raise awareness among individuals in rural communities about the advantages of connecting their homes to electricity. This approach focused on explaining the sustainability, affordability, and tariff plans (payment amounts) associated with having an electricity connection. The findings contribute to the broader discourse on renewable energy development in remote regions, highlighting the significance of tailored solutions that integrate technical expertise with community-driven approaches. Through strategic site optimization and inclusive decision-making, the transition toward sustainable energy in PNG can be accelerated, fostering economic development and environmental stewardship in PNG and beyond.

## REFERENCES

Carravetta, A., Del Giudice, G., Fecarotta, O., & Ramos, H. M. (2013). Pump as turbine (PAT) design in water distribution network by system effectiveness. Water, 5, 1211–1225. https://doi.org/10.3390/w5031211

Cüceloğlu, G., Seker, D. Z., Tanik, A., & Öztürk, İ. (2021). Analyzing effects of two different land use datasets on hydrological simulations by using SWAT model. International Journal of Environment and Geoinformatics, 8(2), 172–185.

Kaur, T., & Segal, R., (2016). Designing rural electrification solutions considering hybrid energy systems for Papua New Guinea Electrical Engineering Department, PEC University of Technology, Chandigarh and 160012, India b GE (Energy Consulting) Bengaluru, India, 1st International Conference on Energy and Power, ICEP2016, 14–16 December 2016, RMIT University, Melbourne, Australia.

Kouadio, K. C. A., Amoussou, E., Diedhiou, A., Coulibaly, T. J. H., Coulibaly, H., Didi, R., & Savané, I., (2022). Using of hydrological model and geospatial tool to assess climate change impact on the hydropower potential of the White Bandama watershed in Côte d'Ivoire (West Africa), IAHS-AISH Scientific Assembly 2022, Montpellier, France, IAHS2022-274, https://doi.org/10.5194/iahs2022-274, 2022.

Nerini, F. F., Broad, O., Mentis, D., Welsch, M., Bazilian, M., & Howells, M. (2016). A cost comparison of technology approaches for improving access to electricity services. Energy, 95, 255–265.

Neves, R., & Leal, V. (2010). Energy sustainability indicators for local energy planning: Review of current practices and derivation of a new framework. Renewable and Sustainable Energy Reviews, 14(9), 2723–2735.

Paish, O. (2002). Small hydro power: Technology and current status. Renewable and Sustainable Energy Review, 6, 537–556.

PNG National Energy Policy (NEP) (2017). Harnessing Energy for life. Department of Petroleum and Energy, Papua New Guinea. ISBN: 978-9950-909-84-8

Rushman, J. F., Thanarak, P., Artkla, S., & Maneechot, P. (2019). Electrical power dem and assessment of a rural community and the forecast of demand growth for rural electrification in Ghana. International Energy Journal, 19, 177–188.

Sammartano, V., Liuzzo, L., & Freni, G. (2019). Identification of potential locations for run-of-river hydropower plants using a GIS-based procedure. Energies, 12, 3446. https://doi.org/10.3390/en12183446

Sekac, T., Jana, S. K., Sutherland, M., & Samanta, S. (2021). Spatio-temporal assessment of rainfall variability and trends in the highland to coastal regions of Papua New Guinea. International Journal of Geoinformatics. http://dx.doi.org/10.52939/ijg.v17i3.1893

Singh, D., (2009). Micro hydro power resource assessment handbook prepared for APCTT Asian and Pacific Centre for transfer of technology of the United Nations – Economic and Social Commission for Asia and the Pacific (ESCAP).

Strauch, M., Bernhofer, C., Koide, S., Volk, M., Lorz, C., & Makeschin, F. (2012). Using precipitation data ensemble for uncertainty analysis in SWAT streamflow simulation. Journal of Hydrology, 414, 413–424, 2012.

# 10 Role of Smart Sensor in Internet of Things for Structural Health Monitoring of Composite Structures

*Alak Kumar Patra and Arun Kumar Singh*

## 10.1 INTRODUCTION

Modern technological and engineering advances include the Internet of Things (IoT) [1] which has dire need for composite structures' structural health monitoring (SHM) using smart sensors [2]. Let's dissect each idea to understand how they work together. The term "IoT" describes a network of physical objects, including cars, appliances, and other things, that are implanted with sensors, software, electronics, and connectivity to allow them to communicate and share data [1]. IoT is important to SHM [3,4] because it makes it possible for different parts of the monitoring system to communicate with each other without interruption. Important elements of IoT in SHM [5–7] comprise devices (sensors or actuators), connectivity (wireless or wired), and data processing platforms (cloud or local servers). Engineered materials composed of two or more phases [8] of materials are known as composite materials. Because of their excellent strength-to-weight ratio and resistance to corrosion, they are frequently used in the maritime, automotive, aerospace, and construction industries. Composite structures are structures made of composite materials. The goal of SHM [9] is to keep an eye on the state of structures in order to identify damage, evaluate its extent, and project how long they will last. Conventional SHM techniques frequently only offer sporadic snapshots of structural health and can be expensive and time-consuming [6,10]. Smart sensors that are integrated into composite structures keep an eye on variables, including moisture content, temperature, vibration, and strain, and give data on structural performance in real time so that damage or anomalies can be found early [11]. Throughout the structure, a number of sensors work together to establish a network that wirelessly transmits data to a cloud-based platform or central processing unit. For effective and safe data transfer, protocols like MQTT, CoAP, or HTTP are used [12]. Cloud computing is used by IoT platforms [13,14] to store, analyze, and visualize data, to identify trends, anticipate possible malfunctions, and enhance maintenance plans; algorithms examine sensor data. Checking for wear and strain on the fuselage and wings of aircraft or



DOI: 10.1201/9781032683928-10

keeping an eye out for cracks, deformations, or structural flaws in buildings, bridges, and other infrastructure is possible with these cohesive systems. The use of smart sensors and IoT-enabled SHM to monitor and control the health of composite structures [15,16] is a paradigm change. Industries can improve safety, extend the life of structures, and maximize maintenance efforts by utilizing real-time data and sophisticated analytics. IoT technologies have the potential to significantly increase the robustness, efficiency, and dependability of composite structures [17,18] in a variety of industries while also establishing new benchmarks for SHM and maintenance procedures.

When it comes to the IoT and composite structures' SHM, smart sensors are essential [17,19]. These are a few of the important functions they play: data acquisition (DAQ) and sensing, real-time monitoring, early damage detection, condition assessment, integration with IoT platforms, cost-effectiveness, customization, and scalability. In this context the role of IoT in SHM of composite structures with smart sensors will be discussed in the following sections.

## 10.2 DATA ACQUISITION AND SENSING

A variety of sensing devices, including strain gauges, temperature sensors, accelerometers, and others, are built into smart sensors, enabling them to directly detect physical characteristics that are vital to the stability of composite structures. They keep gathering information about environmental factors and structural reactions. The overall procedure can be addressed under three key aspects, **data acquisition, sensing, integration and application**. An integrated image of the whole process is shown in Figure 10.1. DAQ as well as the sensing are the primary two operations under this topic. At present, another operational stage is important, that is,

**FIGURE 10.1** Data acquisition, sensing, integration and application. (Courtesy: Free image from stock cake: https://stockcake.com/i/drone-field-research_1004492_835361.)

integration and application which indicates a cohesive unit with multiple sensors and instruments in which data is collected, processed, and transmitted for applications. In this way, integration and application comes under DAQ and sensing. These are briefly presented below for further explanation on this aspect.

## 10.2.1 Data Acquisition

Obtaining data, DAQ, is the process of taking samples of signals that represent physical situations as they exist in the actual world and turning them into digital values that a computer can work with. The typical components involved in this process usually consist of sensors (used to detect physical quantities), DAQ systems that oversee the sampling and digitization processes, signal conditioning (used to transform analog signals into digital form), and analog-to-digital converters (ADCs).

### 10.2.1.1 Sensing

Sensing is the use of sensors to identify the physical characteristics of an environment or an object. Numerous variables, including light, motion, pressure, temperature, and more, can be measured using sensors. Different types of sensors are used to achieve these various types of objectives. More specifically, following are their types along with their applications.

### 10.2.1.2 Types of Sensors

Depending on what they are used to detect, there are numerous types of sensors, including proximity, pressure, temperature, and motion sensors.

Different types of sensors are the key components in sensing operational systems which are presented in more detail as follows.

#### *10.2.1.2.1 Motion Sensors*

As a vital component of SHM, motion sensors identify and quantify vibrations and motions in structures such as buildings, bridges, dams, and other infrastructure. Motion sensors are used in SHM in the following ways.

*10.2.1.2.1.1 Monitoring Structural Integrity* Dynamic behavior of structures is continuously monitored by motion sensors. They pick up vibrations from internal sources like structural deterioration or deformation as well as external ones like wind, traffic, or seismic activity.

*10.2.1.2.1.2 Motion Sensor Types Used*

*10.2.1.2.1.2.1* Accelerometers Accelerometers are multi-axis sensors that measure accelerations. They are frequently employed to keep an eye on a structure's vibrations and dynamic reactions.

*10.2.1.2.1.2.2* Strain Gauges Although their primary use is to measure strain, strain gauges can also be used to indirectly detect motions and vibrations in structures.

*10.2.1.2.1.2.3* Inclinometer Using an inclinometer to measure the tilt or inclination of structural parts can reveal settlements or deformations in the structure.

*10.2.1.2.1.2.4* Laser Doppler Vibrometers Accurate, non-contact sensors monitor surface vibrations; excellent for in-depth dynamic investigation of certain structural locations.

*10.2.1.2.1.3 Uses of Motion Sensors for SHM* *Damage Detection*: Motion sensors can assist in the detection of structural damage or deterioration by tracking variations in vibration patterns over time.

*Health Monitoring*: Ongoing observation makes it possible to evaluate the structural health in real time and spot possible problems before they get out of hand.

*Assessment of Performance*: Information from sensors is used to assess how well structural systems work under various loads and environmental circumstances.

### 10.2.1.3 Analyzing and Interpreting Data

*Analysis of Frequency*: Fourier transforms and other spectrum analysis methods are frequently used to examine motion sensor data in order to pinpoint prominent frequencies and vibration modes.

*Pattern Recognition*: Deviations from normal vibration patterns or frequency spectra may point to damage or structural irregularities.

*Comparative Analysis*: By comparing present conditions with historical data and baseline measurements, one can evaluate how structural health has changed over time.

*Connectivity with SHM Systems Integration*: Larger SHM systems with data gathering devices, processing software, and communication modules often incorporate motion sensors.

They might be networked to give structural engineers access to real-time data, facilitating prompt maintenance scheduling and decision-making.

All things considered, motion sensors offer insightful information on the dynamic behavior and structural health of buildings, which helps to ensure the safer and more effective management of infrastructure assets over time.

#### *10.2.1.3.1 Proximity Sensors*

In order to monitor the state and integrity of items or structural components, proximity sensors are essential to SHM. They do this by detecting and measuring an object's or component's closeness. The usual application of proximity sensors in SHM is as follows.

*10.2.1.3.1.1 Displacement Detection* Proximity sensors are able to pick up on minute motions or displacements within structures. They are employed in the measurement of the separation between components or between structural parts and their environment. This information is useful for evaluating the stability of the structure and identifying any unusual motions that might point to future structural breakdown.

*10.2.1.3.1.2 Vibration Monitoring* Structure vibrations can be observed using proximity sensors. Vibration pattern variations may be a sign of wear and tear, damage, or modifications to the structural dynamics. Engineers can evaluate the condition of the structure and identify any anomalies that could need more research by continuously monitoring vibrations.

*10.2.1.3.1.3 Alignment and Positioning* Proximity sensors are occasionally employed to make sure that structural components are positioned or aligned correctly. In order to prevent structural problems during the construction or maintenance stages, they are essential for detecting deviations from the desired position or alignment.

*10.2.1.3.1.4 Material Integrity* Proximity sensors that rely on eddy currents or electromagnetic induction, for example, are capable of determining the material integrity. They are able to identify alterations in the characteristics of the materials or the existence of flaws like corrosion or cracks below the surface of buildings.

*10.2.1.3.1.5 Real-Time Monitoring* Data on the state of structures is provided in real time by proximity sensors. Through ongoing observation, engineers are able to spot patterns and trends that may point to gradual degradation or damage over time. It makes it possible for prompt maintenance and intervention to stop disastrous failures.

*10.2.1.3.1.6 Integration with IoT and Data Analytics* To facilitate remote monitoring and predictive maintenance, proximity sensors can be combined with IoT platforms and data analytics tools. Sensor data can be used to improve structural performance overall, optimize maintenance schedules, and forecast future failures.

*10.2.1.3.1.7 Use in a Variety of Structures* Due to its adaptability, proximity sensors can be used in a wide range of structures, such as pipelines, buildings, bridges, dams, and industrial facilities. They improve longevity and safety by offering insightful information about the condition and functionality of vital infrastructure.

To sum up, proximity sensors are crucial instruments for tracking the structural health of buildings. They give vital information that engineers can use to evaluate the structural integrity of buildings, spot damage or deterioration early on, and plan maintenance schedules that will maintain infrastructure's long-term performance and safety.

##### 10.2.1.3.2 Temperature Sensors

Temperature sensors are essential to SHM because they provide useful information for evaluating the structural soundness of dams, buildings, bridges, and other infrastructure. The following is the usage of temperature sensors in SHM.

*10.2.1.3.2.1 Measuring Thermal Expansion and Contraction* Materials expand and contract in response to temperature variations. Ongoing observation of these alterations can identify internal strains in the structure that may eventually cause deformation or failure.

*10.2.1.3.2.2 Finding Hotspots* Variations in temperature locally may point to possible problems, such as overheating from mechanical friction, electrical problems, or insufficient airflow.

*10.2.1.3.2.3 Monitoring Environmental Conditions* The long-term durability of a structure can be impacted by many environmental conditions, including sunshine, wind, and humidity, which can be evaluated through the use of temperature data.

*10.2.1.3.2.4 Structural Model Calibration* Temperature measurements are necessary for calibrating simulations and structural models. They offer empirical evidence to support theoretical predictions regarding the behavior of materials at various temperatures.

*10.2.1.3.2.5 Early Warning of Structural Problems* Unexpected or sudden temperature swings can occasionally be a sign of structural issues like corrosion, water intrusion, or insulation failures.

*10.2.1.3.2.6 Integration with Other Sensors* To provide a more thorough knowledge of structural behavior, temperature data frequently supplements data from other sensors (such as strain gauges, accelerometers, or humidity sensors).

*10.2.1.3.2.7 Remote Temperature Sensors* By enabling continuous monitoring without the need for human involvement, remote temperature sensors facilitate early anomaly detection and prompt maintenance interventions.

*10.2.1.3.2.8 Evaluation of Long-Term Performance* By monitoring temperature trends over several months or years, engineers can evaluate the structure's long-term health and performance and schedule maintenance or upgrades appropriately.

In conclusion, temperature sensors in SHM play a critical role in the overall cost-effectiveness, safety, and dependability of infrastructure by supplying essential information for risk management and proactive maintenance.

##### 10.2.1.3.3 Pressure Sensors

Pressure sensors are crucial parts of SHM systems because they provide useful information that engineers can use to evaluate the functionality and state of different types of structures. The following are some important functions and uses for pressure sensors in SHM.

*10.2.1.3.3.1 Monitoring Structural Loads* Pressure sensors are used to monitor structural loads. These sensors pick up forces acting on a structure, such as wind, fluid pressure (like water in dams), or ground pressure (like in foundation monitoring). This data can point to possible overloading or structural flaws and aid engineers in understanding how external loads affect the structure.

*10.2.1.3.3.2 Identifying Structural Deformations* Shifts in pressure may be a sign of structural movements or deformations, such as soil consolidation, foundation settling, or tilting of the structure. Pressure sensors aid in the early detection of these alterations, enabling prompt treatments to stop additional harm.

*10.2.1.3.3.3 Monitoring Fluid Levels and Flow* Pressure sensors keep an eye on hydraulic pressures, fluid levels, and flow rates in structures like dams, bridges, and pipelines. When evaluating the operational safety and structural integrity of such infrastructure, this information is essential.

*10.2.1.3.3.4 Finding Leaks* Pressure sensors are able to identify anomalous drops in pressure, which could point to breaches or leaks in sealed systems like storage tanks, pipelines, or pressurized containers. Early detection aids in avoiding operating disruptions and environmental harm.

*10.2.1.3.3.5 Evaluating Structural Performance under Dynamic Loads* Dynamic pressure changes brought on by industrial vibrations, traffic, or seismic activity are measured by pressure sensors. This information is useful for assessing the structure's resistance to changing loads and for performance-enhancing design optimization.

*10.2.1.3.3.6 Integration with Structural Models* To verify load assumptions, improve predictive models, and optimize structural designs for efficiency and safety, pressure data from sensors is incorporated into structural models and simulations.

*10.2.1.3.3.7 Early Warning Systems* Proactive maintenance and safety actions are made possible by real-time pressure monitoring, which offers early warning of any structural failures or dangerous circumstances.

*10.2.1.3.3.8 Remote Monitoring and Data Logging* Wireless or remote monitoring capabilities for pressure sensors allow for continuous data gathering and logging, which makes it possible to analyze and make decisions in real time without requiring in-person inspections.

Summarily, pressure sensors in SHM systems are essential for maintaining the strength, longevity, and safety of structures because they offer useful information on loads, deformations, fluid dynamics, and operating circumstances. They are essential to contemporary infrastructure management plans because they allow for preventive maintenance and lower the chance of structural breakdowns.

#### 10.2.1.4 Applications

Sensors find usage in a wide range of fields, such as consumer electronics (such as smartphones), automotive systems (such as collision detection), healthcare (such as medical monitoring devices), and environmental monitoring (such as weather stations).

Finally, integration and the applications are presented as an essential stage in modern days for obtaining data which follows.

### 10.2.2 Integration and Applications

#### 10.2.2.1 Integration

Contemporary data collection systems combine a number of sensors and instruments to form unified entities with the ability to gather, process, and send data instantly.

#### 10.2.2.2 Applications

Information obtained from sensors and data gathering systems is used for analysis, monitoring, control systems, and decision-making in a variety of sectors, from industrial manufacturing (e.g., quality control) to scientific research (e.g., physics experiments).

In addition to the DAQ and sensing, smart sensors with IoT capabilities help in real-time monitoring of composite structures' health. This function of IoT is addressed in the following section.

## 10.3 MONITORING IN REAL TIME

Smart sensors with IoT capabilities can send data to cloud- or centralized-based systems in real time. This eliminates the need for physical inspections and enables ongoing monitoring of the performance and structural integrity of composite constructions.

Real-time SHM is made possible by smart sensors that are connected to the IoT. This allows for the continuous gathering, analysis, and preventive maintenance of data on infrastructure, including buildings, bridges, and other vital assets. This is how they collaborate.

### 10.3.1 Sensor Technology

A variety of sensors, including strain gauges, temperature sensors, humidity sensors, and accelerometers, are built into smart sensors. These sensors continuously gather information about the behavior of the structure, including temperature fluctuations, strain levels, and vibrations, all of which are important markers of the structure's health.

Using smart sensors for real-time SHM of composite constructions is an inventive way to guarantee the durability, dependability, and safety of these cutting-edge materials. Composite structures provide special challenges because of their complicated failure mechanisms and vulnerability to damage from a variety of sources, including collisions, fatigue, and environmental variables. Composite structures are frequently utilized in aerospace, automotive, marine, and civil engineering applications.

#### 10.3.1.1 SHM's Smart Sensors: Smart Sensor Types

##### *10.3.1.1.1 Fiber Optic Sensors*

These are sensors with microprocessors which measure temperature, tension, and occasionally damage using optical fibers that are bonded to or implanted in the composite structure.

##### *10.3.1.1.2 Piezoelectric Sensors*

These are used to identify strain and locate and identify damage (such as delaminations) by converting mechanical stress into electrical signals.

##### *10.3.1.1.3 Sensors that Measure Electromechanical Impedance (EMI)*

These sensors use variations in electrical impedance to detect damage-related alterations in structural characteristics.

#### 10.3.1.2 Usability

##### *10.3.1.2.1 Real-Time Monitoring*

Intelligent sensors track changes that could be signs of deterioration or damage by continuously monitoring the strain, temperature, and vibration of the structure.

##### *10.3.1.2.2 Data Acquisition*

For analysis, they gather and send data to a cloud-based system or central processing unit.

*Damage Detection*: SHM systems are able to discover and identify damage by employing sophisticated algorithms or by comparing variations in sensor readings to baseline data.

*Condition Assessment:* SHM offers information about the structure's present state, enabling prompt maintenance or repair measures.

#### 10.3.1.3 Advantages

*Early Detection*: Increases safety and dependability by allowing prompt intervention before damage reaches catastrophic levels.

*Cost-effectiveness*: Lowers maintenance expenses by allowing for targeted fixes and cutting down on downtime.

*Enhanced Performance*: Increases the lifespan of composite constructions and maximizes operational efficiency.

*Integration*: May be retrofitted into already-existing structures or incorporated into the design stage.

#### 10.3.1.4 Problems

##### *10.3.1.4.1 Integrating Sensors*

Making sure they are compatible with composite materials and do not jeopardize structural integrity.

##### *10.3.1.4.2 Data Interpretation*

Creating precise algorithms to analyze sensor data and differentiate environmental effects from damage.

##### *10.3.1.4.3 Reliability*

Ensuring that sensors function consistently in challenging environmental circumstances, such as those seen in aerospace and maritime applications, is known as reliability.

*10.3.1.4.4 Cost*

Long-term advantages may outweigh greater setup and maintenance costs in the short term.

#### 10.3.1.5 Uses

*10.3.1.5.1 Aerospace*

Checking for fatigue fractures or impact damage on the rotor blades, fuselage, and wings of aircraft.

*10.3.1.5.2 Automobiles*

Checking for fatigue or impact damage on carbon fiber reinforced plastic (CFRP) parts.

*Marine*: Checking for corrosion and wear on hulls and important structural elements.

*Civil engineering*: Structural integrity monitoring of wind turbine blades, buildings, and bridges.

In summary, for composite structures, real-time SHM with smart sensors presents a number of benefits in terms of cost-effectiveness, efficiency, and safety. Sustained progress in sensor technology and data analytics is essential to enhance the dependability and versatility of these systems across diverse industries.

## 10.4 EARLY DAMAGE DETECTION

Anomalies or departures from normal behavior can be found early on by examining data from smart sensors. By anticipating possible malfunctions or damages before they worsen, early detection improves safety and lowers the need for expensive repairs.

One of the most important features of SHM that improves the longevity, dependability, and safety of infrastructure is early damage detection via smart sensors. Here's how intelligent sensors help with this procedure.

### 10.4.1 Constant Monitoring

Real-time data on temperature, strain, vibrations, corrosion levels, and other factors is gathered by smart sensors, which allow for the continuous monitoring of structures. The ability to notice minute changes that can point to the beginning of damage or degeneration is made possible by this ongoing monitoring.

### 10.4.2 Data Analysis and Pattern Recognition

Sophisticated algorithms analyze the information gathered by intelligent sensors to find trends or deviations that can point to possible harm. Large datasets can be analyzed using machine learning techniques, which can also be used to spot early warning indicators that may not be immediately apparent from a visual assessment alone.

### 10.4.3 Localization of Damage

Intelligent sensors can be used to locate damage inside a structure. Engineers are able to precisely pinpoint the location of damage or stress concentrations by using triangulation to examine data from several sensors positioned throughout the structure.

### 10.4.4 Early Warning Signals

Smart sensors have the ability to send out early warning signals when anomalies or possible damage are identified. By triggering additional inspections or preventive actions before the damage reaches a critical level, these signals help engineers avoid expensive repairs or disastrous breakdowns.

### 10.4.5 Cost-Effectiveness and Safety

Smart sensors help save costs by facilitating early detection, which lessens the need for costly and time-consuming repairs that would be necessary if damage were permitted to continue unchecked. Additionally, they improve safety by offering prompt information that enables preventative maintenance and risk reduction.

### 10.4.6 Long-Term Monitoring and Trend Analysis

By tracking structural behavior and performance over time, smart sensors make trend analysis easier. Engineers can use this longitudinal data to evaluate the efficacy of maintenance plans and make well-informed choices about the longevity and dependability of the building.

In summary, early damage detection with smart sensors in SHM improves maintenance procedures, increases the operational lifespan of structures, and increases infrastructure safety and dependability overall. It is an example of a proactive structural management strategy that makes use of technology to reduce risks and increase productivity.

## 10.5 CONDITION ASSESSMENT

By utilizing gathered data, smart sensors offer insights into the present state of composite constructions. Assessing structural health, identifying maintenance requirements, and maximizing operating effectiveness can all be done with the help of this data.

In order to guarantee the long-term performance and safety of composite structures, condition evaluation utilizing smart sensors in SHM is essential. Because of their excellent strength-to-weight ratio and resistance to corrosion, composite materials, such as CFRP or fiberglass reinforced polymers (FRP), are frequently utilized in aerospace, automotive, marine, and civil engineering applications. The following are some ways that intelligent sensors help with these structures' condition assessment.

### 10.5.1 Monitoring Structural Integrity

Within composite structures, smart sensors continuously track important factors, including stresses, vibrations, temperature, and humidity. Early detection of variations from typical operating circumstances that can point to damage or deterioration is made possible by this real-time monitoring. An example is shown in Figure 10.2 for wind power system.

### 10.5.2 Finding and Identifying Cracks and Delamination

Two prevalent forms of damage in composite constructions are cracks and delamination, or the separation of layers. Changes in strain distribution or acoustic emissions may be signs of the onset and progression of delamination or cracks, and they can be picked up by smart sensors. Early detection facilitates the prompt initiation of maintenance or repairs to stop more harm.

**FIGURE 10.2** Condition assessment. (Courtesy: Free image from stock cake: https://stockcake.com/i/engineer-analyzing-data_531539_931491.)

### 10.5.3 Localized Damage Detection

Engineers can identify localized damage by carefully positioning smart sensors at key locations on the composite structure, such as close to possible stress concentration zones or at interfaces. This specialized monitoring aids in determining the state of particular areas and informing the selection of appropriate restoration techniques.

### 10.5.4 Monitoring Corrosion

Corrosion is a major hazard for composite structures that contain metallic components (such as bolts and fittings). When corrosion levels rise above acceptable bounds, smart sensors fitted with corrosion sensors or probes can track the pace of deterioration and notify maintenance staff. Proactive monitoring aids in averting structural collapses brought on by weakening parts.

### 10.5.5 Monitoring of the Environment

Smart sensors are also capable of keeping an eye on variables, including humidity, temperature, UV exposure, and chemical exposure. The endurance and long-term performance of composite materials can be impacted by certain environmental factors. Engineers can evaluate the environmental influence on the structure and take necessary preventive action by keeping an eye on these aspects.

### 10.5.6 Predictive Maintenance and Data Analytics

The vast amounts of data gathered by smart sensors can be analyzed using sophisticated data analytics methods, such as machine learning algorithms. These algorithms are capable of pattern recognition, failure mode prediction, and maintenance schedule optimization. By reducing downtime and repair expenses, predictive maintenance based on sensor data extends the operational life of composite structures.

### 10.5.7 Integration with Structural Design and Testing

To verify structural performance under simulated operating conditions, smart sensors can be incorporated into the design and testing stages of composite structures. Through this integration, useful data is made available for enhancing design parameters and guaranteeing that structures satisfy performance and safety requirements for the duration of their lives.

In conclusion, smart sensors are essential to the status evaluation of composite structures because they allow for proactive maintenance plans, early damage identification, and ongoing monitoring. In a variety of industrial applications, engineers may improve the efficiency, safety, and dependability of composite structures by utilizing advanced analytics and sensor technology.

## 10.6 INTEGRATION WITH IoT PLATFORMS

In order to enable data gathering, storage, and analysis, smart sensors are frequently integrated with IoT platforms. Interoperability and smooth connectivity with other

**FIGURE 10.3** Integration with IoT platforms. (Courtesy: Free image from stock cake: https://stockcake.com/i/smart-home-concept_219408_40840.)

IoT systems and devices are made possible by this integration. Figure 10.3 presents a structure which represents residential innovation, where houses are equipped with connected devices and systems that can be controlled remotely.

The capabilities of monitoring, analysis, and decision-making processes are improved when IoT platforms are integrated into SHM of composite structures. IoT systems help with the SHM of composite constructions in the following ways.

### 10.6.1 Data Acquisition and Connectivity

Smart sensors implanted in composite structures can be connected and integrated more easily with the help of IoT platforms. These sensors gather information on a number of variables, including temperature, humidity, vibrations, and stresses. IoT makes it possible to easily acquire data from dispersed sensors via wireless networks, guaranteeing ongoing and real-time monitoring.

### 10.6.2 Centralized Data Management and Storage

IoT solutions offer on-premises or cloud-based centralized data storage. Large volumes of sensor data gathered from various composite architectures can now be managed effectively. Comprehensive condition assessment and trend analysis are made possible by the ability of engineers and maintenance teams to access historical and real-time data from any location.

### 10.6.3 Real-Time Monitoring and Alerts

IoT solutions allow structural health metrics to be monitored in real time. These platforms are able to instantly assess incoming sensor data through integration with intelligent algorithms and analytics. They are able to identify irregularities, anticipate any problems, and provide alerts or notifications to maintenance staff so they may act promptly.

### 10.6.4 Flexibility and Scalability

IoT platforms are flexible enough to expand to monitor more structures as needed and scalable enough to incorporate more sensors. Additionally, they are adaptable in that they can accommodate a variety of sensor kinds and communication protocols, guaranteeing compatibility with diverse composite materials and monitoring specifications.

### 10.6.5 Security and Data Privacy

IoT platforms are designed with strong security features to shield private SHM data from hackers or unauthorized access. Sensors, IoT platforms, and end users may transfer data with confidence, thanks to data encryption, authentication methods, and secure connection protocols.

### 10.6.6 Lifetime Monitoring and Performance Optimization

From initial installation to operational use to end-of-life concerns, IoT solutions enable thorough lifetime monitoring of composite structures. Proactive maintenance, performance optimization, and well-informed decision-making about repair, retrofitting, or replacement options are made possible by continuous data-driven insights.

Finally, combining IoT platforms with SHM of composite structures improves the structural performance of the structures over their lifetime, allows for proactive maintenance, and strengthens monitoring capabilities. Engineers can manage composite structures more efficiently, safely, and reliably in a variety of industrial applications by utilizing IoT technologies.

## 10.7 REMOTE ACCESSIBILITY

Stakeholders (engineers, maintenance staff, etc.) can remotely access both historical and real-time data on the composite structures thanks to SHM systems that use smart sensors. Proactive maintenance plans and prompt decision-making are supported by this accessibility.

The capacity to access and manage SHM data and systems from a remote place is known as remote accessibility with smart sensors in composite structures' SHM.

Improving operating efficiency, facilitating prompt decision-making, and lowering maintenance expenses all depend on this characteristic. The following are some ways that smart sensors and remote accessibility improve composite structures' SHM.

### 10.7.1 Real-Time Monitoring

Smart sensors that are integrated into composite structures gather information on a variety of factors, including temperature, humidity, vibrations, and stresses, continuously. Engineers and maintenance staff can remotely monitor these parameters in real time from any location with internet access. They can observe trends, abnormalities, and current situations without physically being at the site of the structure.

### 10.7.2 Early Detection

Early identification of structural problems or anomalies is made possible by remote access to SHM data. When sensors identify anomalies that can point to possible damage or deterioration, engineers can instantly get alerts or notifications via web interfaces or mobile devices. By taking a proactive stance, the likelihood of expensive repairs or structural breakdowns is reduced by enabling prompt intervention and preventative maintenance.

### 10.7.3 Effective Diagnostics and Troubleshooting

Remote accessibility allows engineers to remotely access diagnostic tools and carry out troubleshooting processes in the event of alarms or unexpected readings from sensors. They are able to evaluate sensor data, gauge the seriousness of problems, and choose the best course of action. This feature shortens reaction times and minimizes downtime related to onsite visits or manual checks.

### 10.7.4 Optimized Maintenance Scheduling

Maintenance teams can base their maintenance schedules on real performance data rather than predetermined intervals by remotely accessing historical data and trend analysis reports. By ensuring that maintenance is carried out when required, this data-driven method maximizes the operational lifespan of composite structures and reduces the expense of needless maintenance.

### 10.7.5 Cost-Effective Operations

Manual data collection and recurring onsite inspections are less necessary when accessibility is achieved remotely. This financial advantage is especially noteworthy for sizable or widely distributed composite structures, as it reduces travel costs and downtime related to maintenance tasks.

### 10.7.6 Enhanced Safety and Dependability

The safety and dependability of composite structures are improved by prompt access to SHM data and proactive maintenance procedures facilitated by remote accessibility. Engineers are able to quickly put corrective measures into place, reduce any dangers, and guarantee that structures are kept in top shape for the duration of their working lives.

### 10.7.7 Integration with IoT and Cloud Platforms

Smart sensor integration with cloud-based SHM systems and IoT platforms enables remote accessibility. These solutions provide scalability, data integrity, and security standard compliance while offering safe access to sensor data via web interfaces or mobile applications.

In summary, engineers and maintenance teams may now monitor, evaluate, and manage structural health data remotely thanks to smart sensors in SHM of composite structures. In a variety of industrial applications, it fosters the long-term dependability and safety of composite structures, strengthens operational efficiency, and facilitates proactive decision-making.

## 10.8 COST-EFFECTIVENESS

By decreasing the frequency of physical inspections and offering continuous monitoring, smart sensors provide a more cost-effective alternative than traditional inspection methods, which may incur downtime and labor costs.

Smart sensor IoT devices are a preferred option for enterprises looking for cost-effective maintenance techniques since they provide various advantages over other options when it comes to SHM of composite structures. The following are the main ways that smart sensors improve the cost-effectiveness of composite structures' SHM.

### 10.8.1 Early Detection and Preventive Maintenance

Structural characteristics including temperature, vibrations, and stresses are continuously monitored by smart sensors. They facilitate preventive maintenance interventions by identifying early indicators of damage or degradation. By taking care of problems early on, you may lower the likelihood of catastrophic failures that could cause costly downtime or safety hazards, as well as the repair costs associated with advanced damage.

### 10.8.2 Lower Inspection Costs

Manual labor, specialized equipment, and downtime for structural access are frequently needed for traditional inspection methods. The continuous monitoring capabilities of smart sensors eliminate the need for regular physical examinations. This decrease in the frequency of inspections avoids operational disturbances and reduces personnel expenses, particularly for big or remote composite constructions.

### 10.8.3 Data-Driven Maintenance Planning

An enormous amount of historical and real-time data regarding the performance of the structure is gathered by smart sensors. Maintenance teams can optimize maintenance schedules and prioritize tasks based on actual structural conditions instead of calendar-based intervals by analyzing this data. By minimizing pointless maintenance tasks and optimizing the effectiveness of maintenance efforts, this strategy conserves time and resources.

### 10.8.4 Extended Structural Lifespan

Composite constructions have an extended operating lifespan thanks to proactive monitoring and prompt maintenance made possible by smart sensors. Smart sensors assist in minimizing deterioration over time by resolving problems quickly and maximizing performance. Long-term cost reductions are substantial since the frequency of major repairs or early building replacements is decreased due to this lifespan extension.

### 10.8.5 Remote Monitoring and Management

Engineers and maintenance staff can monitor and manage structures from any location with internet access thanks to smart sensors' ability to make SHM data remotely accessible. This feature allows for faster reaction times to alarms or anomalous readings and lowers the travel expenses related to onsite inspections. Through the ability to make decisions and take action more quickly, remote monitoring also improves operational efficiency.

#### 10.8.5.1 Flexibility and Scalability

IoT-based smart sensor systems are flexible enough to add more sensors or increase their monitoring capacities as needed. Because of its scalability, SHM solutions can expand or alter the infrastructure without needing to replace or adapt expensive monitoring systems.

Predictive maintenance solutions are made possible through the integration of smart sensors and predictive analytics tools. Predictive analytics can anticipate possible problems and suggest preventive maintenance procedures by examining patterns and trends in sensor data. By minimizing unscheduled downtime and emergency repairs, this predictive method maximizes operational effectiveness while lowering overall maintenance costs.

In conclusion, by facilitating early detection, data-driven decision-making, remote monitoring, and predictive maintenance, smart sensors as IoT devices in SHM of composite structures offer considerable cost-effective advantages. These advantages improve the longevity, safety, and dependability of composite structures in a range of industrial applications in addition to lowering operating costs.

## 10.9 IMPLICATIONS FOR ACADEMICIANS AND IMPLICATIONS FOR PRACTITIONERS

Though implications are discussed in all the sections above, starting from the last part of Section 10.1 through Section 10.8, some specific implications for the practitioners and the academicians are presented in the following sub-sections.

### 10.9.1 Customization and Scalability

Smart sensors with IoT capabilities can be tailored to meet the unique monitoring requirements of various composite structures. Additionally, they are scalable, enabling the expansion of monitoring capabilities through the addition of sensors as required.

In order to improve flexibility and scalability in composite structures using smart sensors for SHM, the IoT is essential. Here's how IoT makes these things possible.

#### 10.9.1.1 Personalization of Surveillance Systems: Sensor Selection and Configuration

The IoT provides a broad selection of smart sensors suited to particular monitoring requirements, allowing for the customization of SHM systems. Depending on the features and weaknesses of the composite construction, engineers can choose sensors based on factors like strain, vibration, temperature, and corrosion.

*Integration with Current Infrastructure*: IoT systems provide the integration of smart sensors with current infrastructure for monitoring. By adding sensors to various composite constructions, this integration enables customization for applications in civil, automotive, aerospace, and marine engineering.

#### 10.9.1.2 Real-Time Gathering and Interpretation of Data

*Constant Monitoring*: Real-time data collection is made possible by IoT-enabled smart sensors, which are capable of continuously monitoring structural factors. This continuous data stream makes it possible to identify anomalies or departures from typical operating conditions quickly, which enables timely intervention and preventive maintenance.

*Data Fusion and Analysis*: IoT systems make it easier to combine and evaluate data from many sources and sensors. This data is processed using advanced analytics methods, like machine learning algorithms, to find trends, correlations, and possible problems in composite constructions. Engineers are able to extract practical knowledge that may be applied to maximize efficiency and prolong the life of structures.

#### 10.9.1.3 Scalability in Various Applications

*Flexible Deployment*: IoT-based SHM systems can be extended and adjusted to fit a range of composite structures and uses. IoT enables the deployment of sensors in a scalable way, supporting various sizes, types, and environments, whether monitoring a single structure or a network of structures.

#### 10.9.1.4 Extension Capabilities

By introducing new sensors or modernizing old ones, IoT enables the scalability of SHM systems as infrastructure develops or grows. Because of their scalability, monitoring solutions may expand to keep up with the development of infrastructure without needing significant retrofitting or replacement.

#### 10.9.1.5 Remote Control and Accessibility

*Remote Monitoring*: Engineers and stakeholders can monitor composite structures from any location with internet connectivity thanks to IoT's ability to provide remote access to SHM data. This capacity facilitates prompt decision-making, alert response, and effective structural health management.

#### 10.9.1.6 Centralized Control

Data management and monitoring systems are centralized under IoT platforms. By conducting diagnostics, adjusting monitoring parameters, and configuring sensors remotely, engineers can improve operating efficiency and lower maintenance costs related to onsite inspections.

#### 10.9.1.7 Improved Dependability and Safety

*Proactive Maintenance*: IoT supports proactive maintenance tactics by enabling scalability, customization, and real-time monitoring. Based on data-driven insights, engineers may put preventive measures into place to reduce hazards and guarantee the longevity of composite constructions' safety and dependability.

#### 10.9.1.8 Predictive Analytics

By integrating IoT, predictive analytics may be used to anticipate possible problems and improve maintenance plans. IoT data-derived predictive insights reduce downtime, prevent expensive repairs, and extend the useful life of composite constructions.

To sum up, IoT allows for customized monitoring solutions, real-time data gathering and analysis, flexible deployment across applications, remote accessibility, and proactive maintenance techniques, all of which improve customization and scalability in SHM of composite structures. These skills enable engineers to manage composite structures in a variety of industrial sectors cost-effectively, while also maximizing performance and ensuring safety.

### 10.9.2 Challenges and Emerging Trends

Several new trends, obstacles, and conclusions are presented by the application of the IoT to SHM of composite structures using smart sensors.

#### 10.9.2.1 Emerging Trends

IoT device integration with smart sensor integration for real-time data collection and analysis in composite structure SHM: This is a developing trend. Early detection of structural abnormalities and ongoing monitoring are made possible by this integration.

#### 10.9.2.2 Wireless Sensor Networks

Large-scale sensor networks can be installed within composite constructions thanks to developments in wireless communication technologies. These networks make it possible to transmit data across large distances without a lot of wiring.

#### 10.9.2.3 Data Analytics and Artificial Intelligence (AI)

Using data gathered from IoT sensors, machine learning algorithms and artificial intelligence (AI) are used for anomaly detection and predictive maintenance. The accuracy and effectiveness of SHM systems are being improved by this trend.

Edge computing is the practice of processing data close to the sensors, at the network's edge, as opposed to sending all of the raw data to a single, central server. Edge computing speeds up decision-making by lowering latency and bandwidth consumption.

#### 10.9.2.4 Remote Monitoring and Control

Composite constructions can be remotely monitored and controlled with IoT-enabled SHM systems. This feature is very helpful for buildings in dangerous or isolated areas.

#### 10.9.2.5 Standardization and Protocols

To guarantee interoperability and ease of integration in SHM applications, efforts are being made to define standardized protocols and interfaces for IoT devices and smart sensors.

#### 10.9.2.6 Challenges/Problems

##### *10.9.2.6.1 Data Security*

One of the biggest challenges facing IoT devices is ensuring the security and integrity of the data that is transmitted and stored by them. IoT network vulnerabilities could jeopardize the confidentiality and dependability of SHM data.

##### *10.9.2.6.2 Power Management*

Since remote or inaccessible locations are common for IoT devices and sensors in SHM applications, power management and energy efficiency are essential for long-term operation.

##### *10.9.2.6.3 Sensor Accuracy and Calibration*

The efficacy of SHM systems depends on the smart sensors' capacity to remain accurate and reliable over time. It's essential to calibrate and monitor sensor health on a regular basis.

##### *10.9.2.6.4 Scalability*

Careful design of sensor placement, network architecture, and data management procedures are necessary for scaling up IoT-enabled SHM systems to monitor big or complex structures.

#### *10.9.2.6.5 Cost*

The initial setup expenses might be high and include the purchase of sensors, IoT devices, and infrastructure. The investment in IoT for SHM needs to be justified by cost-effective solutions.

#### *10.9.2.6.6 Interoperability*

For smooth integration and operation of SHM systems, it is essential to ensure interoperability among various IoT devices and sensors from different manufacturers.

## 10.10 CONCLUSION

In conclusion, the use of IoT to smart sensor-enabled composite structures' SHM has the potential to completely transform the way these structures are maintained and observed. SHM systems can offer real-time insights about structural health by utilizing IoT technologies, such as edge computing, data analytics, and wireless sensor networks. This allows for preventive maintenance and lowers the chance of catastrophic failures. The insights developed from the chapter facilitate achieving sustainable development goals by leveraging AI and IoT. By utilizing smart sensors and IoT-based SHM, the infrastructures will last longer by reducing greenhouse gas emission from new constructions, and conversion of more lands to human infrastructures. Additionally, it will enable access to modern technology and infrastructures to everyone, thereby facilitating SDG 9: Industry, Innovation and Infrastructure of the United Nations (UN). Moreover, this will enhance the achievement of goal 11—sustainable cities and communities—as well as goal 12—responsible consumption and production of the UN due to real-time data transmission, analysis, and informed economic decision-making for longer life of composite structures. But in order to fully utilize IoT in SHM, issues including data security, sensor dependability, scalability, and affordability must be resolved. In order to overcome these obstacles and propel the adoption of IoT-enabled SHM solutions in the future, technological developments must continue, as must industry collaboration and standardization initiatives.

## REFERENCES


1. Li, S., Xu, L. D., & Zhao, S. (2015). The internet of things: A survey. *Information Systems Frontiers*, *17*, 243–259.
2. Azad, M. M., Kim, S., Cheon, Y. B., & Kim, H. S. (2023). Intelligent structural health monitoring of composite structures using machine learning, deep learning, and transfer learning: A review. *Advanced Composite Materials*, *33*(2), 162–188. https://doi.org/10.1080/09243046.2023.2215474.
3. Tokognon, C. A., Gao, B., Tian, G. Y., & Yan, Y. (2017). Structural health monitoring framework based on Internet of things: A survey. *IEEE Internet of Things Journal*, *4*(3), 619–635.
4. Mishra, M., Lourenço, P. B., & Ramana, G. V. (2022). Structural health monitoring of civil engineering structures by using the internet of things: A review. *Journal of Building Engineering*, *48*, 103954.
5. Weihnacht, B., & Tschöke, K. (2021). Smart monitoring and SHM. In *Handbook of Nondestructive Evaluation 4.0* (pp. 1–16). Cham: Springer International Publishing.

6. Saravanan, T. J., Mishra, M., Aherwar, A. D., & Lourenço, P. B. (2024). Internet of things (IoT)-based structural health monitoring of laboratory-scale civil engineering structures. *Innovative Infrastructure Solutions*, *9*(4), 110.
7. Scuro, C., Sciammarella, P. F., Lamonaca, F., Olivito, R. S., & Carni, D. L. (2018). IoT for structural health monitoring. *IEEE Instrumentation & Measurement Magazine*, *21*(6), 4–14.
8. Yao, J., Zhou, Z., Zhou, H., Yao, J., Zhou, Z., & Zhou, H. (2019). Introduction to composite materials. *Highway Engineering Composite Material and Its Application*, 1–23.
9. Guemes, A. (2013, November). SHM technologies and applications in aircraft structures. In *Proceedings of the 5th International Symposium on NDT in Aerospace, Singapore* (Vol. 1315, p. 4).
10. Abdelgawad, A., & Yelamarthi, K. (2017). Internet of things (IoT) platform for structure health monitoring. *Wireless Communications and Mobile Computing*, *2017*(1), 6560797.
11. Wang, J., Fu, Y., & Yang, X. (2017). An integrated system for building structural health monitoring and early warning based on an Internet of things approach. *International Journal of Distributed Sensor Networks*, 13(1), 1–14. doi: 10.1177/1550147716689101
12. Jeong, S., Hou, R., Lynch, J. P., Sohn, H., & Law, K. H. (2019). A scalable cloud-based cyberinfrastructure platform for bridge monitoring. *Structure and Infrastructure Engineering*, *15*(1), 82–102.
13. Botta, A., De Donato, W., Persico, V., & Pescapé, A. (2016). Integration of cloud computing and internet of things: A survey. *Future Generation Computer Systems*, *56*, 684–700.
14. Rao, B. P., Saluia, P., Sharma, N., Mittal, A., & Sharma, S. V. (2012, December). Cloud computing for Internet of Things & sensing based applications. In *2012 Sixth International Conference on Sensing Technology (ICST)* (pp. 374–380). IEEE.
15. Mohapatra, A. G., Talukdar, J., Mishra, T. C., Anand, S., Jaiswal, A., Khanna, A., & Gupta, D. (2022). Fiber Bragg grating sensors driven structural health monitoring by using multimedia-enabled IoT and big data technology. *Multimedia Tools and Applications*, *81*(24), 34573–34593.
16. Rahita, A. C., Zaki, A., Nugroho, G., & Yadi, S. (2024). Internet of things (IoT) in structural health monitoring: A decade of research trends. *Instrumentation, Mesures, Métrologies*, *23*(2), 123–139.
17. Changazi, S. A., Bakhshi, A. D., Yousaf, M., Mohsin, S. M., Akber, S. M. A., Abazeed, M., & Ali, M. (2024). Optimization of network topology robustness in IoTs: A systematic review. *Computer Networks*, *250*, Retreived August 2024 from https://doi.org/10.1016/j.comnet.2024.110568. Retreived August 2024, https://doi.org/10.1016/j.comnet.2024.110568
18. Di Sante, R. (2015). Fibre optic sensors for structural health monitoring of aircraft composite structures: Recent advances and applications. *Sensors*, *15*(8), 18666–18713.
19. Hollaway, L. C. (2010). A review of the present and future utilisation of FRP composites in the civil infrastructure with reference to their important in-service properties. *Construction and Building Materials*, *24*(12), 2419–2445.

# 11 State of Geospatial Data for Environmental Governance in Bangladesh

*Sheikh Tawhidul Islam and Bashir Ahmed*

## 11.1 INTRODUCTION

Economic development in a sustainable manner is emphasized in macro policies and sectoral strategies in Bangladesh. These policy documents provide directions and options for developing necessary institutional makeup, mentioning geospatial data needs to perform activities relating to sustainable development in different sectors. Bangladesh has made a good progress in attaining Millennium Development Goals (MDGs) and now focusing on achieving Sustainable Development Goals (SDG) targets. SDG targets are different from MDG targets in terms of thematic areas, coverage and comprehensiveness. It is widely perceived that one of the most important challenges relating to SDG attainments is to generate required data so that progress made in achieving SDG targets could properly be monitored, shared and reported. In doing that United Nations Statistical Division (UNSD) provided instruments such as Environment Statistics Self-Assessment Tool (ESSAT) to address the data generation, sharing and reporting challenges for member countries, including Bangladesh. In this connection, data generation agencies of Bangladesh such as Bangladesh Bureau of Statistics (BBS) have been trying to align environmental data generation architecture, especially the geospatial data architecture, including related management and reporting systems of Bangladesh, as per the guidelines provided by UNSD.

BBS is the prime and mandated agency in Bangladesh to generate data on various environmental themes. However, many other agencies working in areas like water resources (e.g., Water Resources Planning Organization, WARPO), meteorology (e.g., Bangladesh Meteorological Department, BMD), soil (e.g., Soil Resources Development Institute, SRDI) and agriculture (e.g., Bangladesh Agricultural Research Council, BARC) generate various types of spatial and non-spatial environmental data. Thus, environmental indicators are generated by various agencies as per their requirements adopting different methods; these methodological variations sometimes create obstacles (e.g., incompatibility) to assimilate data when there is a necessity. In addition, data are generated at different scales with limited disaggregation provisions (e.g., geographical disaggregation, gender, and income-based disaggregation). BBS recently developed Bangladesh Environmental Statistics Framework (BESF) 2016–2030 following the guidelines provided by United Nations-Framework for Development of Environment Statistics (UN-FDES) and

DOI: 10.1201/9781032683928-11

United Nations-System of Environmental Economic Accounting (UN-SEEA) so that data gathered by different sources could be used in order to develop data for reporting and monitoring the progress of attaining SDGs. The review of literature and institutional processes on environmental data generation in Bangladesh suggests that lack of institutional commitments in (geospatial) data generation and sharing, required funding, high rate of staff turnover among the institutions, skills and knowledge gaps in data generation methods and techniques still remain as major barriers in aligning existing data generation provisions.

## 11.2 NATIONAL ENVIRONMENTAL CONTEXTS AND DATA GENERATION

Sustenance and security of majority of people, their living arrangements and conditions in Bangladesh have been heavily dependent on the availability, access and quality aspects of natural resources (Ministry of Environment and Forests [MoEF] 2018). These natural resources like (i) land with favourable properties such as soil fertility, moisture holding capacity, well drainage systems, (ii) biological diversity and resources that provide wide array of choices to cultivate crops by the farmers, (iii) water resources that ensure irrigation facilities, supply of important nutrient elements for plants and animals, and (iv) climatic conditions like optimum temperature, occurrence of adequate rainfall, wind direction and flow pattern, necessary sunshine, all in a combined fashion create an enabling environment suitable for natural resource-based primary production systems in this country (Brammer 2014). The food and livelihood security for millions of people directly and indirectly depend on these primary production processes. Impacts on these natural resources and physical systems caused by unsustainable development activities, environmental pollution and degradation, disasters including climate change result in destabilizing the (agricultural) production systems, leading to food insecurity, poverty and creating social tensions (Ministry of Environment and Forests [MoEF] 2017).

This profound dependency on natural resources of people of Bangladesh suggests that sustainable development is the most pragmatic approach to follow in Bangladesh since it will help to uphold environmental integrity and at the same time allow economic development to continue. The macro policy instruments of Bangladesh like the Perspective Plan 2041, 8th Five Year Plan have advocated about sustainable development and mentioned green development plans as most appropriate path while implementing development programmes.

A number of reports were developed by relevant Bangladeshi agencies since 2001 that contain different environmental data to support environmental governance, natural resource planning and management. For instance, BBS plans to develop 15 new sets of reports (following the UN-FDES guidelines) where a number of reports are focusing on environment, disaster and climate change impacts. The first report on the state of environment in Bangladesh was developed and published by Department of Environment (DoE) with support of UNEP in 2001 based on the data that were available up to 1995 for the purpose of developing Global State of the Environmental Report, 2002. The report pulled data from different sources and presented the facts in narrative form about land degradation, water pollution and scarcity, air pollution, biodiversity and natural disasters using the analytical framework known as pressure-state-impact-response (PSIR). The report also indicated that the environment, social and development problems are multifaceted and it requires more complete (not partial), precise and analytical and properly linked/relational data so that human actions and resultant development impacts could be properly understood. It also recommended to create and foster long-term mechanisms to monitor and assess the effects of green policies on environmental quality and quality of life of the human being. The report could be treated as a baseline state of environment in Bangladesh.

The Country Environmental Analysis for Bangladesh was published by Asian Development Bank (ADB) in 2004. The report focused on the environmental concerns especially the pollution in the water and air, discussed about the release of solid, hazardous/toxic chemicals, hospital wastes, arsenic contamination in the ground water, land degradation and soil quality (e.g., salinity intrusion, fertility decline, nutrient imbalance, loss of organic matter), loss of top soil issues etc. The report mentioned about the legal, institutional and policy frameworks relating to environment of Bangladesh. It indicated that about 200 environmental laws are available in Bangladesh.

BBS published Compendium of Environmental Statistics of Bangladesh in 2009 under the institutional capacity development. The report contained data on biodiversity, climatic variables, land and water resources, human settlement, information on wastes and pollution (e.g., air, soil, sound, water and radioactive pollution) and impacts of disaster and climate change. BBS published Disaster-related Statistics in 2015 to capture impacts of natural disasters, including disasters induced from climate change on human lives and livelihoods. The study was focused on 12 major disaster types occurring in different parts of Bangladesh. The report showed that about 56.52% of the households experienced major disasters once in their life, while 26.57% and 16.91% reported their experiences twice and thrice, respectively (BBS 2015). Floods appeared to be the most disturbing and damaging disasters compared to all other disasters as 24.44% of respondents reported that they were affected by floods, followed by 15.10% affected by cyclones and 10.59% mentioned about thunderstorm impacts.

As indicated before, BESF (2016–2030) following the principles and guidelines provided by UN-FDES and SEEA. The objectives of BESF are (i) identifying main quantifiable aspects of the environment, (ii) identifying components, sub-components and topics that are relevant and statistically feasible according to defined national needs and priorities, (iii) facilitating the development of a national programme of environmental statistics, (iv) contributing to the assessment of data requirements, sources, availability and gaps, (v) guiding the development of databases that can be used for multiple purposes and (vi) assisting the coordination and organization of environmental statistics given the inter-institutional nature of the domain. It is indicated

**TABLE 11.1**
**National Reports on Environment-Related Statistics of Bangladesh since 2001 and Published and Proposed Report Titles under BESF 2016–2030**

| No. | Title of Report | Year of Publication | External Support |
|---|---|---|---|
| 1. | State of Environment (SoE) Report | 2001 | UNEP in collaboration with DoE, BCAS, SACEP and NORAD |
| 2. | The Country Environmental Analysis: Bangladesh | 2004 | Asian Development Bank |
| 3. | Bangladesh – Country Environmental Analysis | 2006 | World Bank |
| 4. | Compendium of Environment Statistics of Bangladesh | 2009 | Capacity Building of Bangladesh Bureau of Statistics Project |
| 5. | Disaster-related Statistics 2015 (Climate Change and Natural Disaster Perspectives) | 2015 | Impact of Climate Change on Human Life (ICCHL) Programme |
| 6. | Bangladesh Environmental Statistics Framework (BESF) | 2017 | UNEP and UNDP |
| **Reports to be Developed as Proposed in the BESF (2016–2030)** | | | |
| 1. | Compendium of Environmental Statistics | | |
| 2. | Compilation of Resource Accounts following SEEA (land/soil, water, forests, natural gas, energy, fish) | | |
| 3. | Climate Change and Natural Disaster-related Statistics | | |
| 4. | Compilation of Social Accounting Matrix | | |
| 5. | Poverty Environment Accounts (PEA) in light with SEEA | | |
| 6. | Experimental Ecosystem Accounts (EEA) in light with SEEACF | | |
| 7. | Household Survey of Health and Sanitation in Disaster Prone Areas of Bangladesh | | |
| 8. | Urban/Rural Waste Generation Recycling and Management Survey | | |
| 9. | Environmental Protection and Resource Management Expenditure Accounts | | |
| 10. | Disaster Risk Reduction Expenditure Accounts | | |
| 11. | Climate Change and Natural Disaster Impacts Vulnerability Index | | |
| 12. | Pre-crisis Data Gathering Tools as Baseline Information | | |
| 13. | Climate and Natural Disaster Induces Survey | | |
| 14. | Urban/Rural Water Generation Use and Management Survey | | |
| 15. | Developing Web-Based Data Sharing Reporting and Ensuring Access to Stakeholders | | |

in the BESF that BBS will prepare 15 sets of reports (Table 11.1) by the year 2030. It is imperative to mention that General Economics Division (GED) of Planning Commission also conducted a report on SDG data gap analysis and assessed the availability of indicators with 62 government agencies/Ministries in terms of data readily available, partially available and not available. Development of BESF (2016–2030) consulted all related documents, including this GED report in outlining the proposed title of (15 sets of) reports.

## 11.3 ENVIRONMENTAL GOVERNANCE IN BANGLADESH: POLICIES AND INSTITUTIONAL MANDATES

Almost all sectors within Bangladesh have environmental concerns of some description, especially those relating to action in natural resources management such as land and water. The National Environmental Council was established in Bangladesh, chaired by the prime minister. It functions through an Executive Ministerial Committee headed by the Ministry of Environment Forest and Climate Change (MoEFCC) and a Divisional Environment Committee headed by the Divisional Commissioner. The MoEFCC is primarily responsible for environmental protection, also responsible for the formulation and monitoring of environmental policy and legislation. This ministry acts as the controlling authority of all executing agencies like DoE, Forest Department (FD), Bangladesh Forest Research Institute (BFRI), Bangladesh Forest Industries Development Corporation (BFIDC) and Institute of Forestry and Environmental Sciences (IFESCU). Furthermore, it coordinates other inter-ministerial (e.g., water, industrial, transport and mining) environmental issues. The FD works as an executing agency for the protection, control, conservation, expansion and maintenance of the national forest resources. In addition, a number of Ministries take into consideration environmental issues as cross-cutting. These include Ministry of Planning, Ministry of Local Government Rural Development and Cooperatives (MoLGRD & C), Ministry of Water Resource, Ministry of Agriculture, Ministry of Health and Family Welfare, Department of Public Health Engineering, Water Supply and Sewerage Authority, Ministry of Energy and Mineral Resources.[1] The National Environment Management Action Plan (NEMAP) 1992 of Ministry of Environment and Forest (MoEF) proposes actions to achieve the objective stated in the National Environmental Policy 1992. These actions cover many diverse areas related to the environment with emphasis on public participation in the process of formulating the action plan. The Environmental Conservation Act of 1995 empowered the MoEF to formulate rules and guidelines for the management of environment. It also designates DoE responsible for enforcing the 1997 EIA procedures to control air pollution, water pollution, noise pollution. The EIA process is categorized into four classes, those are green, amber A, amber B and red[2](Table 11.2).

It is important to note that the legal frameworks that are concerned about environment in Bangladesh are crowded with big number of documents. However, the Bangladesh Environment Policy 1992,[3] the Forest Policy 1994 (Forest Act 2021), the Water Policy 1999, the National Land Use Policy 2001, the National Fisheries Policy 1998, the National Environmental Management Action Plan (NEMAP) 1995 are the major policies that play roles in providing necessary guidelines in formulating acts, laws, strategies, plans and to issue relevant ordinances on environmental issues. The policies also provide directives in making institutional architecture and to formulate institutional mandates for achieving the policy objectives. The important issues to mention here are that most of the policies were developed before the year 2000 in the contexts and necessity of that time and hence the institutional architecture, human resources, mandates to achieve were remained to be old fashioned. In contrast, the policies those were formulated or amended after 2000 incorporated the environmental and biodiversity conservation issues more seriously.

**TABLE 11.2**
**Some of the Key Environmental Laws in Bangladesh**

| Laws and Regulations | Relevance to Environment |
|---|---|
| Bangladesh Environmental Conservation Act, 1995 | Empowers the MOEF to formulate rules and guidelines for the management of the environment. Designates DOE as responsible for enforcing the 1997 EIA procedures. |
| Environmental Conservation Rules, 1997 | Air pollution, water pollution, noise |
| Environmental Pollution Control Ordinance, 1997 | National water quality standards according to WHO guidelines, air quality standards, noise, solid waste management |
| Agriculture Pest Ordinance, 1962 | Toxic and hazardous substance |
| Private Forest Ordinance, 1950 | Forest Conservation |
| Forest Act, 1927 | Forest Conservation, biodiversity conservation, soil conservation |
| Wildlife Preservation Act, 1973 | Wildlife conservation, wetland management, biodiversity conservation |
| Wildlife (Conservation and Security) Act 2012 | |
| Medical Waste Management Rules, 2009 | Management of medical wastes generated in hospitals, clinics and diagnostic centres |
| Marine Fisheries Ordinance, 1983 | Coastal resources management, biodiversity conservation, marine pollution |
| Territorial Water and Marine Zone Act, 1974 | Coastal resources management, marine pollution |
| Water Supply and Sewerage Authority Ordinance, 1963 | Water resources management |
| Ship Breaking and Hazardous Waste Management Rules, 2010 | |
| Balu Mohal and Soil Management Rules 2011 | |
| Bangladesh Climate Change Resilience Fund, 2011 | |
| Forest Transit Rule, 2011 | |
| Draft National River Conservation Act, 2011 | |
| Bangladesh Wildlife Conservation and Security Act, 2012 | |
| Draft Tree Conservation Act, 2012 | |
| Disaster Management Act, 2012 | |
| Forest (Amendment) Act, 2012 | |
| Sustainable & Renewable Energy Development Authority (SREDA) Act (draft), 2012. | |
| Land Zoning Act (Draft), 2012 | |
| National Water Act (draft), 2012 | |
| Brick Production Act (draft), 2012 | |
| Haor Master Plan, 2012–2032 | |
| Bangladesh Wildlife Conservation and Security Act, 2012 | |
| Disaster Management Act, 2012 | |
| Forest (Amendment) Act, 2012 | |
| Draft e-Waste Management Rule, 2018 | |

It suggests that the legal documents that were passed earlier were less aligned with the national and international goals and targets relating to environmental conservation, including SDG green targets. This overcrowding and old makeup of the legal documents leave impacts on the ways environmental concerns are to be handled by relevant agencies.

## 11.4 UTILIZATION OF EARTH OBSERVATION AND GEOSPATIAL INFORMATION SYSTEM DATA

The United Nations Statistical Commission (UNSC) and United Nations Committee of Experts on Global Geospatial Information Management (UN-GGIM) have recognized the need for managing and effectively integrating geospatial and statistical information nationally and globally. It is imperative to mention here that application of satellite remote sensing, photogrammetry for analysing aerial photographs have been in use in Bangladesh since 1980 with the development of Space Research and Remote Sensing Organization (SPARRSO) and thereafter emergence and contributions of Environment and GIS (EGIS) in the water sector in 1990. In addition, Survey of Bangladesh (SoB), SRDI, BARC are some agencies who have the capacities to use geospatial methods, data and tools in Bangladesh. SPARRO have been using data from meteorological satellites like National Oceanic and Atmospheric Administration-Advanced Very High Resolution Radiometer (NOAA-AVHRR) to forecast rainfall, tornado and cyclone predictions and the results are supplied to BMD and BBS for their reporting and dissemination purposes. SoB produces administrative and topographic maps at 1:50,000 scale and aerial photographs covering the entire country are archived for years like 1974–1975 (1:30,000 scale), 1983/1984 (1:50,000 scale) and 1999–2000 (at 1:25,000 scale). But use of these data needs permissions from Bangladesh Ministry of Defence. The agency named EGIS, later renamed CEGIS, has necessary expertise to use radar data for flood monitoring; the agency also contributes in areas of floodplain research, river morphology dynamics, land use zoning and environmental impact assessments. In addition, natural resources mapping, monitoring, crop estimation, post-disaster impact assessments are some avenues where these agencies play important roles in generating primary data. Bangladesh has got the necessary capacity at this moment to generate and handle space and GIS-based data that may combine disaster data along with disaggregated socio-economic data for evidence-based policy making. Tables 11.3 and 11.4 show the usage type of applied GIS data by SoB.

Tables 11.3 and 11.4 show that there is a large amount of geospatial data available in Bangladesh through various public and private sector organizations which were produced to facilitate their mandated responsibilities to perform. Almost every year the demand for maps and geospatial data is growing in Bangladesh due to the increasing number of prominent governmental organizations within Bangladesh using GIS facilities in their respective fields of application, though most of these users are developing GIS databases in isolation. This is creating redundancy, inconsistency and duplication of data along with very high initial overhead costs.[4] According to the Statistics Act 2013, BBS is mandated to establish an Integrated Geographical Information System. BBS is now working closely with SoB and other GIS-based organizations towards establishing an integrated geographical information system.

**TABLE 11.3**
**Availability of Geospatial Data That Is Collected by SoB**

| | Scale of the Maps: 5,000 and 25,000 (Being Implemented) | |
|---|---|---|
| | **Category** | **Summary Layers** |
| 1 | Administrative Boundary | International, Division, District, Upazila along with Pillars and topographical sheet boundaries |
| 2 | Building and Structure | Building, Building Rooftops, Clustered Buildings, Built up Area, Monuments etc. |
| 3 | Facilities | Religious, Education, Health, Governmental facilities etc. |
| 4 | Geodetic Control Points | Nationwide Vertical and Horizontal control points |
| 5 | Hydrography | River, Wetland, Island/Char-land etc. |
| 6 | Industry | Major Industrial locations along with type |
| 7 | Relief | Contour, Spot heights etc. |
| 8 | Transportation | Road, Railroad, Bridges, Ports etc. |
| 9 | Vegetation | Forest, Cultivation and non-cultivation area |

**TABLE 11.4**
**Usage of Geospatial Data That Is Collected by SoB**

| Field of Applications | Use Activities |
|---|---|
| Agriculture | Monitoring, evaluation and management |
| Environment | Monitoring, modelling and management for land degradation; weather and climate modelling, prediction and forecasting; river and coastal erosion modelling; flood management etc. |
| Health | Aerial distribution of different diseases in relation to environmental factors; visualizing changes in the occurrence of a disease over time etc. |
| Intelligence | Monitoring, tracking, evaluation and management |
| Forestry | Management, planning; map prepare for site specific matching etc. |
| Regional/Local Planning | Development of plans, maintenance, management; infrastructure development programme, land registration etc. |
| Research and Education | Different solutions from personal to national level etc. |
| Resource | Management, planning, monitoring, recording etc. |
| Social Studies | Demographic trends and developments analysis etc. |
| Transport Network | Planning and management etc. |
| Military Use | IPB, DMP and other battle planning, terrain analysis, resource planning, deployment, management, monitoring, recording etc. |
| Other Uses | Thematic mapping, topographical mapping, site and location information, services, consultancy etc. |

Furthermore, within the National Strategy for Development of Statistics (NSDS), scopes have been identified for modernizing GIS and developing a web-enabled GIS mapping system with other statistical attributes.[5]

## 11.5 QUALITY ASSURANCE AND VALIDATION METHODS

Data quality concerns arise in a number of ways in Bangladesh. These quality issues may be elaborated firstly as flaws and shortcomings of data collection methodologies, secondly as introduction of human errors during data collection processes and finally as lack of regular update of the data that may cause loss of relevance of use of data. Data on demographic (through census surveys), agricultural crops, socio-economic indicators and sector-specific variables (e.g., health and sanitation, industrial growth and poverty incidence) are being gathered by BBS since the independence of Bangladesh. In conducting census surveys (e.g., population, agriculture), BBS uses door-to-door data collection methods, while other surveys like Household Income and Expenditure Survey (HIES[6]) the agency is based on representative sample surveys for data collection through a number of Primary Sample Units (PSU). This indicates that census survey results may supply information up to the lowest tier of the country but HIES may be only applicable to district level (not very effective to use lower than district level) reporting. BBS provides training at two levels (develop master trainers who train enumerators and data entry operators) in order to reduce errors and to ensure quality in producing data.

Agencies such as SoB, SRDI, Department of Land Records and Survey (DLRS) also produced data on land ownership and land quality aspects. But due to lack of data update, some of the data become irreverent and thus provide confusing state of the phenomenon. For instance, data generated on different characteristics of soil generated in the 1960s are still in use in making the agro-ecological database (AEZ database) of Bangladesh and these datasets are still in use for agricultural crop production planning and harvest estimation. Karim and Iqbal (2001) showed that soil properties in different AEZs of Bangladesh have significantly changed (Figure 11.1) and commented that without update of the soil data and associated AEZ database the use of this data for agricultural planning might have little use/impact. Later, during the 1990s, the use of geospatial data and techniques and relevant data by a number of Bangladeshi agencies added new dimension in the integration of non-spatial data with non-spatial variables and to perform different kinds of spatial analysis and mapping activities. Data produced on environmental indicators (e.g., waterbodies and forests) by the then ESPAN (currently known as CEGIS), AEZ database including maps developed by BARC, pilot projects implemented by DLRS, infrastructure and service facilities data produced by Local Government and Engineering Department (LGED) are some examples of production of geospatial data in Bangladesh. But these datasets (both spatial and non-spatial) were produced few decades back and become less relevant nowadays because the landscapes of Bangladesh (e.g., geomorphological settings and waterbodies/rivers) are highly dynamic and land configurations change quickly which cannot be captured in databases without regular updating provisions.

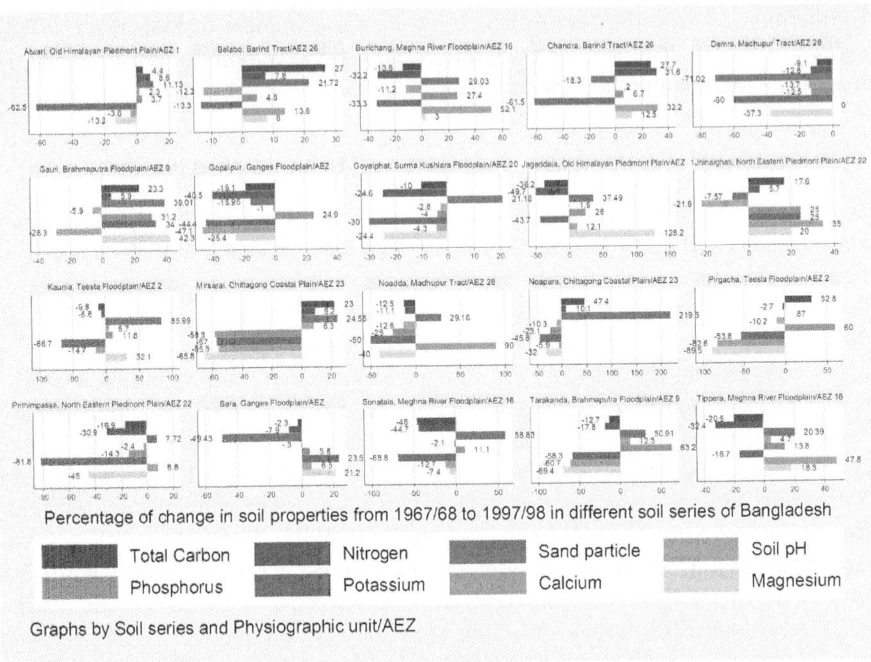


**FIGURE 11.1** Change in the soil properties in different Agro-Ecological Zones of Bangladesh. (Figure generated using data from Karim and Iqbal 2001.)

## 11.6 AVAILABILITY OF ENVIRONMENTAL DATASET BY OTHERS (I.E. DATA USERS)

In general, the agencies which produce environmental data (i.e. act as producer) also use the same data (i.e. act as user) for public reporting, planning and management of natural resources and in some cases for providing early disaster warning to people. For instance, BMD produces data on different meteorological indicators based on daily records, satellite information and old records for climate trend analysis and at the same time uses their own information for long-term forecasts and trend analysis. BMD uses satellite data such as National Oceanic and Atmospheric Administration-Advanced Very High Resolution Radiometer (NOAA-AVHRR) and other satellites to predict rainfall and cyclone warnings. Bangladesh FD also works both as data producer and user. But agencies like Flood Forecasting and Warning Centre (FFWC) of Bangladesh Water Development Board (BWDB) do not produce primary data but use hydro-meteorological and land morphological data generated by BWDB and BMD and also use satellite altimeter data to forecast flood warning signals for various stakeholders like communities, navigating vessels and disaster impact professionals and volunteers. Other agencies such as Department of Disaster Management (DDM), Bangladesh Red Crescent Society, development partners such as UNDP, UNEP, FAO use environmental data generated by different government agencies and help various agencies by developing partnership programmes so that they can provide more effective services to the communities. Programmes like Comprehensive Disaster Management Programme (CDMP,[7] Phases I and II), running from 2003 to 2015, and National Alliance for Risk reduction and Response Initiatives (NARRI)[8], running since 2010, are a few major initiatives implemented in Bangladesh that were supported by a consortium of donors. These programmes use various kinds of environmental data and provide backstopping support to frontline government agencies for their capacity development and enable them to provide community support effectively. DDM with the support of development partners conducted a disaster risk assessment called Community Risk Assessment (CRA) that performs nine-step exercise using different kinds of environmental data and finally prepares Risk Reduction Action Plan (RRAP) at grassroots levels. In addition, this DDM agency produces post-disaster impact assessment data using a field level data collection tool called "D Form" which contains 27 types of loss and damage information categories (Table 11.5) to be filled in by government representatives at upazila level and district commissioners at district level and sent to Emergency Operation Center (EOC) within three weeks of disaster occurrence. The EOC then compiles all the data gathered from the field and develops national loss and damage database and forward to National Disaster Response Coordination Center (NDRCC) based at the Ministry of Disaster Management and Relief (MoDMR) for wider dissemination.

## 11.7 CHALLENGES AND CONSTRAINTS FOR ENVIRONMENTAL DATA MANAGEMENT

The challenges and constraints that exist within the various ministries and agencies for collecting and managing environmental statistics generally entail a shortage of knowledgeable and competently trained human resources coupled with a lack of technical capacity and severe limitation on financial resources. More specifically, the key challenges for having an effective environmental data management and reporting system among the various stakeholders include the following[9]:

- Lack of inter-ministerial/agencies agreement and coordination on how to work together and collect, manage and share data. Ministries, institutions and organizations have various types of environmental data, but due to lack of coordination and formal agreements, it is not possible to access or use this data strategically;
- Lack of a standard or common format and sharing/dissemination platform for administrative environmental data;
- No data quality assurance/guideline is currently available and as a result agencies are reluctant to ensure quality of data they supply or use;
- Lack of permanent and sufficient budget has become one of the vital issues for data collection, compilation, processing and dissemination of environmental data; and
- Absence or non-operationalization of designated ministry, institute or organization focal points that have the official responsibility for management environmental statistical data and other information.

**TABLE 11.5**
**Loss and Damage Information Categories Contained in the "D" Form**

| Information Category | Description | Information Category | Description |
|---|---|---|---|
| 1 | Name of upazila and district affected disasters | 15 | Damage of mobile phone towers |
| 2 | Number of wards/unions affected | 16 | Damage of structures of religious institutions |
| 3 | Affected area in square kilometres | 17 | Information on the damage of road networks of different categories |
| 4 | Affected people (men, women, children) | 18 | Number of bridge and culvert damage |
| 5 | Physically challenged persons (men, women, children) | 19 | Damage of embankments in kilometres |
| 6 | Affected households (partial, total) | 20 | Affected forest areas in hectares |
| 7 | Number of affected house (concrete, semi-concrete, thatch made) | 21 | Number of affected educational institutions |
| 8 | Affected disaster shelters (partial, total) | 22 | Affected industries (agriculture and non-agriculture) |
| 9 | Value of livestock lost (goats, lambs) | 23 | Number of affected tubewells |
| 10 | Value of livestock lost (cow, buffalo) | 24 | Affected toilets/latrines |
| 11 | Value of birds/poultry lost (chicken, duck) | 25 | Affected water reservoirs in numbers |
| 12 | Affected crops and seedbeds in hectares | 26 | Affected health centres (hospitals, clinics, community health centres) |
| 13 | Damage of other farms (e.g., shrimp hatchery) | 27 | Loss of fishing boats and gears (boats, trawlers, fishing nets) |
| 14 | Damage of power lines (partial, total) | | |

*Note:* This form is used by the Department of Disaster Management (DDM) for collecting loss and damage data of disaster impacts.

## 11.8 CONCLUSIONS AND WAY FORWARD

Despite having a good number of governance instruments such as policies, laws, acts, rules, environmental degradation could not be halted as expected. In addition, rapid and slow-onset disasters in the contexts of climate change continue posing threats to physical systems and processes (Islam and Neelim 2010). It is important to note here that at-scale, updated data may help to understand the situations properly and identify the drivers of change on time, which may facilitate taking appropriate action plans to arrest further degradation. Bangladesh Government has adopted UN-ESSAT, UN-FDES, UN-SEEA to outline her environmental data generation process and utilize geospatial tools and methods towards effective governance so that primary production presses are supported by the natural resource base and people can live in a better environmental state with good health. Finally, a number of recommendations are made to strengthen environmental data generation capacity of the agencies/professionals and to develop better management, sharing and reporting systems in Bangladesh. These are as follows: (i) Institutional commitments are necessary to align data generation activities and improve data sharing mechanisms (e.g., assigning officials for data management, disseminating data via web portals). (ii) Inter-agency coordination must be efficient, effective. (iii) Data sharing must be free of costs (since data were generated using public resources) so that sharing is free of hindrance. NSDS 2013, Statistical Act 2013, National Spatial Data Infrastructure (NSDI) should be properly harmonized with UN-FDES and UN-SEEA. In addition, data generation should also be aligned with DRSF (disaster statistics proposed by UNESCAP) and poverty environment integration (PEI) protocols so that overlaps in data generation could be avoided. (iv) Data quality concerns should be addressed by identifying the sources of error (e.g., shortcomings of data collection methodologies and use of technologies to reduce human error). Metadata[10] should be maintained for all kinds of data generated. (v) Data collection strategies/methodologies should be developed as per the requirement of scale so that necessary elements could be taken into consideration (at-scale) for data generation. (vi) Updating existing data is essential since old data could mislead the progress monitoring and planning provisions. (vii) New data acquisition methods, such as satellite remote sensing, should be incorporated in the data generation process. In addition, data (as long as possible) generated for different variables should be geo-referenced and presented and archived in geocoded mapping framework so that environmental variables could be distributed spatially and spatial analysis could be performed.

## NOTES

1. *Handbook on National Environmental Legislation and Institutions in Bangladesh.*
2. *Handbook on National Environmental Legislation and Institutions in Bangladesh.*
3. Bangladesh Environment Policy has been updated; National Environment Committee has approved the updated Bangladesh Environmental Policy, 2018.
4. Bangladesh Country Report Integrating Statistical and Geospatial Information System.
5. Bangladesh Country Report Integrating Statistical and Geospatial Information System.
6. HIES 2016 was conducted comprising 2304 Primary Sample Units (PSUs) covering 46,080 households, while HIES 2010 took only 12,240 households.
7. http://www.bd.undp.org/content/bangladesh/en/home/operations/projects/All_Closed_Projects/Closed_Projects_Crisis_Prevention_and_Recovery/comprehensive-disaster-management-programme/CDMPHome.html
8. http://narri-bd.org/
9. Bangladesh Environmental Statistics Framework (BESF) 2016-2030, May 2017.
10. Data about data.

## REFERENCES

Brammer, H. (2014) Climate Change Sea Level Rise and Development in Bangladesh. University Press Limited, Dhaka.

BBS (2015) Impacts of Climate Change on Human Lives, Ministry of Planning, Government of Bangladesh, Dhaka.

Ministry of Environment and Forests (MoEF) (2017) Climate Investment Program for Environment Climate and Climate Change, Government of Bangladesh, Dhaka.

Ministry of Environment and Forests (MoEF) (2018) Third National Communication, Government of Bangladesh, Dhaka.

Islam, S.T. and Neelim, A. (2010) Climate Change in Bangladesh: A Close Look into Temperature and Rainfall Data, University Press Limited (UPL), Dhaka.

Karim, Z. and Iqbal, M. A. (2001). Impact of Land Degradation in Bangladesh: Changing Scenario in Agricultural Land Use. Soils Division, Bangladesh Agricultural Research Council (BARC), Dhaka.

# 12 Establishing a New Sustainable Rice Cultivar in the Cyclone-Affected Zones of the Sundarbans Mangrove Ecosystem

*Sunanda Batabyal, Nilanjan Das, Sohini Gangopadhyay, Rituparna Banerjee, Subodh Chandra Pal, and Sudipto Mandal*

## 12.1 INTRODUCTION

Tropical cyclones have been more common in coastal India recently, with cyclones like Aila (May, 2009), Fani (April, 2019), 'Bulbul' (November, 2019), 'Umphun' (May, 2020), 'Yaas' (May, 2021), causing destruction. Recent research Chowdhury et al. (2023) showed samples of water from nearby rivers, including the 'Bidya' 'Matla' Hooghly' and 'Raimangal' indicated higher saline levels in the Sundarbans. In a different study, Tenhunen et al. (2023) found that post-Aila conditions led to a significant increase in saline levels measured in the tidal waters of the Sundarbans.

The level of surface waters are elevating which may be contributed by the startling speed of sea level rise in the Sundarbans, which is predicted to be 3.5 mm year$^{-1}$, as well as the frequent occurrence of tropical cyclones like 'Aila', 'Umphun', Yaas' and a very recent Sitrang in October 2022 caused inundation of saline water into the agricultural lands, leading to thousand hectares of land which are unusable and unsuited for agriculture (Mandal et al., 2024; Roy and Ghosh, 2024). The fragility of the rice crop (*Oryza sativa* L.) to saline stress, which is a major contributor to the region's agricultural economy, exacerbates the issue for the rural population.

Rise of soil salinity affects crop yield and lead to food emergencies (Allakhverdiev et al., 1999; Egea et al., 2023). By 2050, soil salinity could wipe out 50% of all arable lands worldwide, as per the 'Food and Agricultural Organization (FAO)' (FAO, 2017; Negacz et al., 2022). A total of 86 million hectares of land in India have been damaged by salinity (Mishra et al., 2023; Pathak, 2000). Rice production in the area can be exceedingly difficult to increase or even maintain when under salt stress. Rice production remains natural till the salt level reaches 3 dS m$^{-1}$, beyond this level, the production decreases nearly 13% for every unit rise in salinity (Omar et al., 2023). The area's main crop, rice, is hindered by water which enters the arable lands with tides, which causes the 'Kharif' (monsoon) growing season plagued by water

DOI: 10.1201/9781032683928-12

logging, whereas due to the capillary rise of 'groundwater' sources within paddy field plots, the Boro (winter) crop is hindered by significant salt stress.

The most crucial component of a salty environment is NaCl. High levels of NaCl have been linked to a variety of negative effects, including an imbalance in the uptake of $K^+$ and $Ca^{2+}$ by rice plants, the inactivation of enzymes, the inhibition of protein synthesis, the premature senescence of leaves, lower photosynthetic and respiration rates, and higher levels of reactive oxygen species (ROS) (; Munns, 2002). Other effects include nitrogen metabolism, availability of water, ion uptake (Guo et al., 2022). Salinity also contributes to a number of germination issues with seeds and propagules (Laamari et al., 2024).

The lipid peroxidation process, which is triggered by stress-related high ROS levels and causes DNA and protein damage, is extremely detrimental to plant cells and destroys membrane lipids (Basu et al., 2010). The production of ROS and/or the suppression of the system that protects against them are thought to be at least largely responsible for the deleterious consequences of these numerous environmental stresses (Bor et al., 2003; Gul et al., 2022). Most abiotic stresses, such as chilling (Wu et al., 2022), heavy metal toxicity (Rashid et al., 2023), drought (Pamungkas and Farid, 2022), and salt (Ondrasek et al., 2022), appear to result in an increase in ROS, which are also created in the course of aerobic metabolism. The defence mechanisms of plants include release of enzymes like catalase (EC 1.11.1.6), glutathione reductase (EC 1.6.4.2), ascorbate peroxidase (EC 1.11.1.11), and glutathione reductase (EC 1.15.1.1), as well as hydrophilic and hydrophobic antioxidants ($\alpha$-tocopherol, carotenoids), ascorbic acid, and glutathione (Sachdev et al., 2023). In order to maintain ROS equilibrium, plants generate the antioxidant enzymes 'SOD' and 'CAT', but 'proline' helps plants to combat with 'osmotic stress' (Upadhyay et al., 2011). A more effective antioxidative system is typically associated with greater resistance to environmental challenges (Singh et al., 2022). When salt concentrations or temperatures are unfavourable, osmo-protectants increase cytoplasmic 'osmotic pressure' and perform a vital function in 'protein' and 'membrane' 'stabilization' (Choudhary et al., 2022; McNeil et al., 1999). Osmo-protectants come in three major varieties: betaines and related compounds, 'polyols and sugars', and amino acids like 'proline'. Plants under salt or water stress produce more 'proline', which has been associated with a species' ability to withstand osmotic stress (Rhodes et al., 1998; Seleiman et al., 2022). Although some researchers (Awaad, 2023; Cao et al., 2023) believe in its importance in fostering resistance to salt stress is debatable; it may be an indication of stress injury rather than a measurement of stress tolerance. Consequently, a deeper analysis of the function of 'proline' build-up in salt stress is needed. According to Dubey and Rani (1989), rice is a crop with a moderate salt tolerance that comes in a variety of salinity-sensitive cultivars.

Salinity is a serious issue that not only lowers agricultural potential but also has an impact on farmer livelihood methods in Sundarbans region (Mistry, 2022). The production of native rice varieties has been negatively impacted by increased soil salinity, and there has been a move towards salt-tolerant cultivars and practises as well (Mistry, 2022). So, here the question arises how does salt stress influence stress enzyme activity and proline content in rice seedlings in a coastal environment? To solve the research question, the current study sought to examine the capacities of

Nona Bokra and IET 4786 (control) to manage the impact of 'salt stress' on their germination rates, antioxidant enzyme activities, and build-up of 'proline'.

## 12.2 MATERIALS AND METHODS

### 12.2.1 Study Site and Seed Collection

Seeds from var. 'IET 4786' and var. '*Nona Bokra*' were considered for the experiment. For the collection of seeds, the present research focused on 'Crop Research and Seed Multiplication Farm (CRSMF)', (Lat: 23°15′4.09″N, Lon: 87°50′41.05″E) of the District of Purba Burdwan and Maipith (Lat: 21°51′58.37″N, Lon: 88°30′53.34″E) of the District of South 24 Parganas, Sundarbans of West Bengal state, India (Figure 12.1). The seeds of rice variety 'IET 4786' were collected from CRSMF and seeds of *Nona Bokra* were collected from Maipith by the local farmers. The salinity of soil in the CRSMF was 2.45 g $l^{-1}$, while in Maipith farming areas, it was 14.78 g $l^{-1}$.

'*Nona Bokra*' is one of the widely used, high-yielding and saline-tolerant varieties that have been found in India so far and suggested for site-specific cultivation in saline environments (Jayabalan et al., 2022). Following a 24-hour period of sterilization with '0.05% prochloraz emulsifiable' concentrate, seeds were thoroughly rinsed with tap water to eliminate the pesticide (Demiral and Türkan, 2005).

### 12.2.2 Experimental Design and Study of Seed Germination

To assess the NaCl stress on germination, in a Whatman No. 2 filter paper, 20 seeds were distributed from each cultivar in 10 cm 'petri dishes'. To them, '5 ml' of either

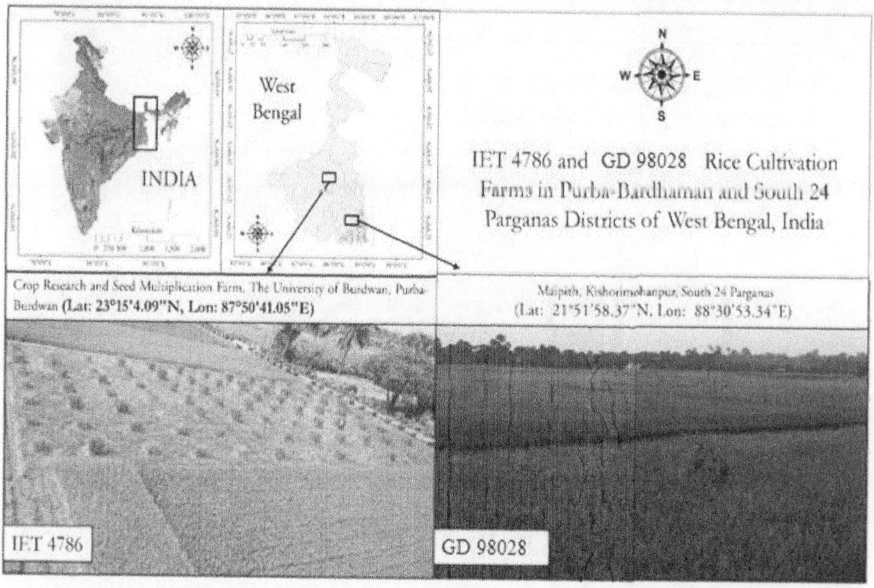


**FIGURE 12.1** Study sites of var. 'IET 4786' and var. 'GD 98028' rice varieties.

'distilled water' (used as a control) or a NaCl solution were added. The seeds were kept in environmental chamber at 30°C with 12 hours of light for 72 hours. A volume of '5 ml' of 'distilled water' (control) or a NaCl solution (60 and 120 mol $m^{-3}$) were added to the plates after two, three, and four days. For six days, the number of seeds that germinated was counted every day.

### 12.2.3 NaCl Stress Treatments of Seedlings

The seedlings were then studied to test how the NaCl stress limits seedling growth. In the regulated 'environmental chamber', the 16-hour photoperiod was maintained at 30°C during day and 22°C night, and 70% relative humidity. This was done using a metal halide bulb. The investigation was carried out for a further 15 days. The nutritional solution comprised 0.88 mol $m^{-3}$ $K_2SO_4$, 1 mol $m^{-3}$ $Ca(NO_3)_2$, 1 mol $m^{-3}$ $(NH_4)_2SO_4$, 1 mol $m^{-3}$ $MgSO_4$, 0.25 mol $m^{-3}$ $KHPO_4$, 0.1 mol $m^{-3}$ KCl, 40 mol $m^{-3}$ Fe-EDTA, 10 mol $m^{-3}$ $H_3BO_4$, 1 mol $m^{-3}$ $MnSO_4$, 1 mol $m^{-3}$ $ZnSO_4$, 0.1 mol $m^{-3}$ $CuSO_4$, and 0.01 mol $m^{-3}$ KCl $(NH_4)_6MoO_{24}$ (Zeng and Shannon, 2000). This solution was changed every two days, and 1 N HCl was used to adjust the pH to 6. The plants were salinized with 0, 60, or 120 mol $m^{-3}$ NaCl nutrient solution up to 15 days. At 8, 11, and 14 days after being exposed to the salt, samples of the leaves and roots were taken (Figure 12.2). Leaf samples were collected, weighed, and stored at freezer to be examined for

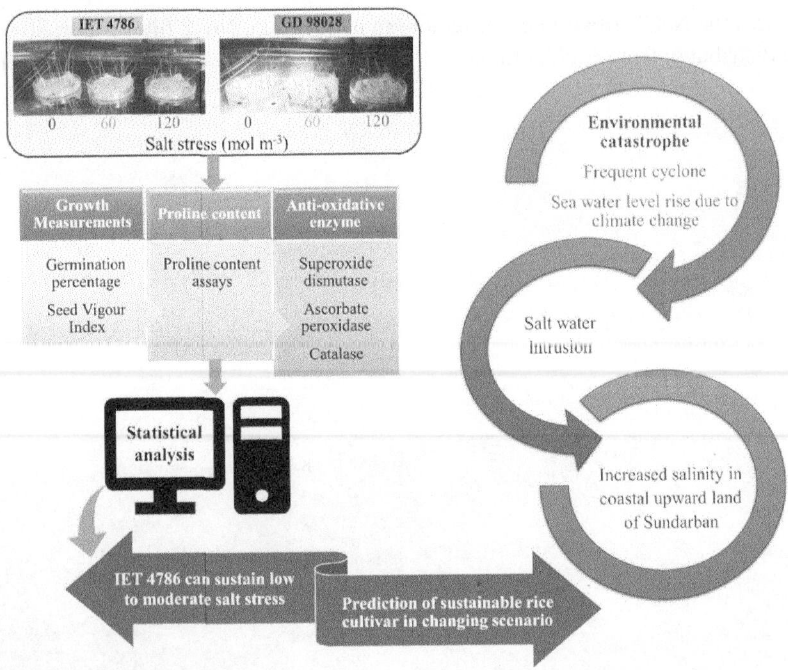


**FIGURE 12.2** Control and salinity stress condition after eight days of seed germination in var. 'IET 4786' and var. 'GD 98028' rice varieties.

antioxidant enzyme activity and 'proline'. Each experiment was independently carried out at least three times.

### 12.2.4 Growth Measurements

For a randomly chosen sample of salt-treated rice seedlings, growth ratios were computed. Following blotting, fresh weights and root lengths were calculated by counting the intervals between the start and end of the NaCl treatment.

### 12.2.5 Antioxidative Enzyme and 'Proline' Content Assays

For the assay of antioxidative enzyme and 'proline' content, 0.5 g sample of leaves were frozen using liquid nitrogen and ground finely. Superoxide dismutase ('SOD', EC 1.15.1.1) activity was tested following the method by McCord and Fridovich (1969). To ascertain the APX activity, the rate of ascorbate oxidation at 290 nm was measured (Nakano and Asada, 1981). The activity of catalase ('CAT', EC 1.11.1.6) was measured by observing the drop in absorbance at 240 nm caused by $H_2O_2$ oxidation (Blume and McClure, 1980). The 'proline' content was determined using the protocol.

### 12.2.6 Statistical Analysis

The collected data of different enzymes were subjected to two-way analysis of variance (ANOVA) using statistical analysis package: SPSS (Version23).

## 12.3 RESULTS

Without being subjected to NaCl stress, 'IET 4786' rice seeds germinated more quickly than '*Nona Bokra*' (Figure 12.3), the germination rate for both the cultivars slow down with the rise of NaCl concentration. In contrast to '*Nona Bokra*', 'IET 4786' had a stronger inhibitory impact. For instance, 'Nona Bokra' germination rate was remained at 80% after six days of salt stress, in the presence of 120 mol $m^{-3}$ NaCl, however, for 'IET 4786', it was only about 10%, respectively, in 120 mol $m^{-3}$ NaCl.

'SOD' activity increased gradually as salinity increased in '*Nona Bokra*' as well as in 'IET 4786'. But '*Nona Bokra*' showed much higher values compared to 'IET 4786' ($p < 0.05$). Additionally, the basal levels of 'SOD' activity of salt-tolerant cultivar '*Nona Bokra*' was substantially higher than that of salt-sensitive cultivar 'IET 4786'. The values were 69.7 ± 0.35, 70.37 ± 0.95, 94.68 ± 0.31 and 95.23 ± 0.64 μmol $min^{-1}$ $mg^{-1}$ protein of fresh weight at the end of 14 days experiment in case of 60 and 120 mol $m^{-3}$ NaCl treatment, for 'IET 4786' and '*Nona Bokra*', respectively (Figure 12.4).

Similar to 'SOD', in the leaves of unstressed plants, 'CAT' activities were higher in that of tolerant cultivar '*Nona Bokra*' 1.89 ± 0.02 and 1.95 ± 0.01 μmol $min^{-1}$ $mg^{-1}$ protein of fresh weight than it was in the salt-sensitive IET 4786, 1.58 ± 0.02 and

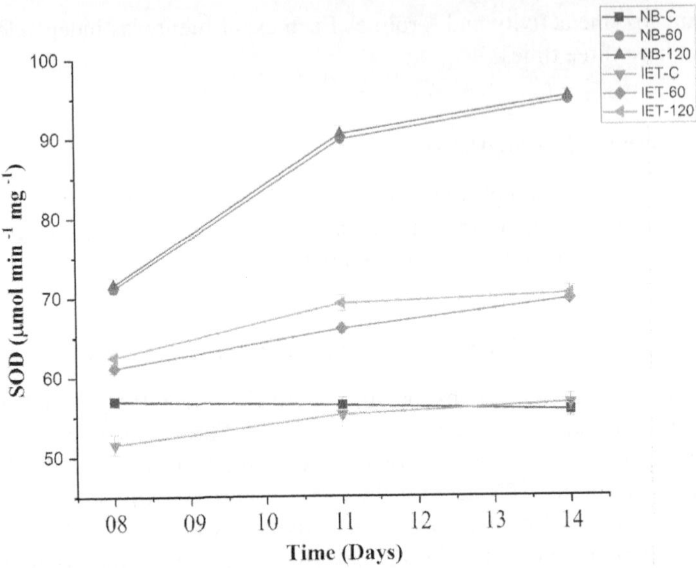


**FIGURE 12.3** Superoxide dismutase (SOD) activity of var. 'IET 4786' and var. 'GD 98028'. Here NB-C = GD 98028 at control condition, NB-60 = GD 98028 at 60 mol m$^{-3}$ NaCl, NB-120 = GD 98028 at 120 mol m$^{-3}$ NaCl, IET C = IET 4786 at control, IET 60 = IET 4786 at 60 mol m$^{-3}$ NaCl, IET 120 = IET 4786 at 120 mol m$^{-3}$ NaCl.

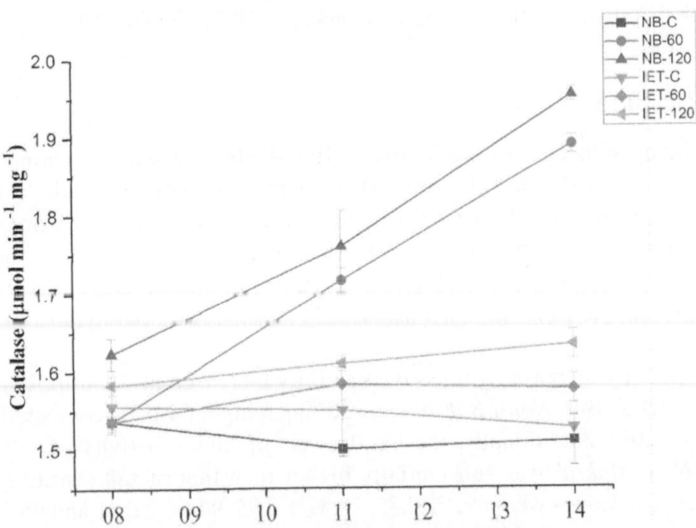


**FIGURE 12.4** Catalase (CAT) activity of var. 'IET 4786' and var. 'GD 98028'. Here NB-C = GD 98028 at control condition, NB-60 = GD 98028 at 60 mol m$^{-3}$ NaCl, NB-120 = GD 98028 at 120 mol m$^{-3}$ NaCl, IET C = IET 4786 at control, IET 60 = IET 4786 at 60 mol m$^{-3}$ NaCl, IET 120 = IET 4786 at 120 mol m$^{-3}$ NaCl.

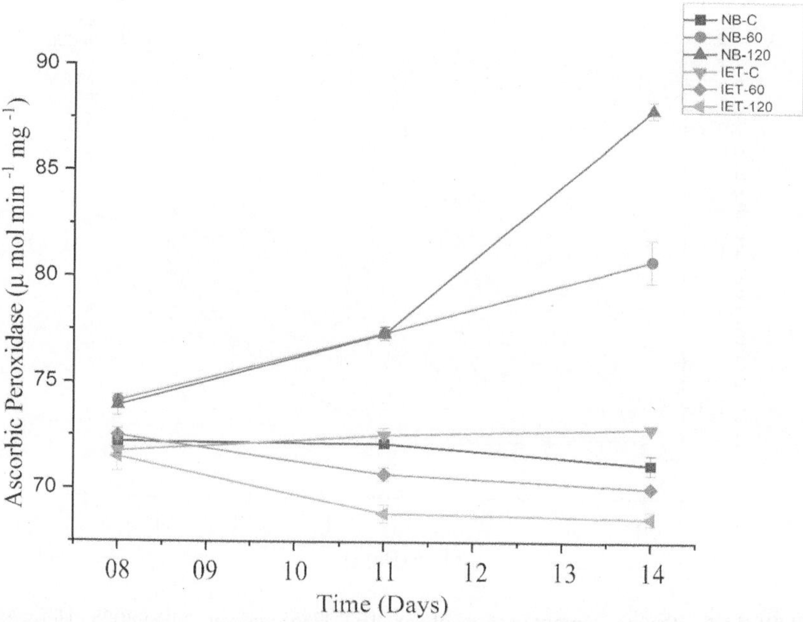


**FIGURE 12.5** Ascorbic peroxidase (APX) activity of var. 'IET 4786' and var. 'GD 98028'. Here, NB-C = GD 98028 at control condition, NB-60 = GD 98028 at 60 mol $m^{-3}$ NaCl, NB-120 = GD 98028 at 120 mol $m^{-3}$ NaCl, IET C = IET 4786 at control, IET 60 = IET 4786 at 60 mol $m^{-3}$ NaCl, IET 120 = IET 4786 at 120 mol $m^{-3}$ NaCl.

1.63 ± 0.02 μmol $min^{-1}$ $mg^{-1}$ protein of fresh weight at the end of 14 days experiment in 60 mol $m^{-3}$ and 120 mol $m^{-3}$ NaCl treatment (Figure 12.5).

In contrast to the salt-sensitive cultivar 'IET 4786', salt-tolerant cultivar '*Nona Bokra*' showed significantly increasing levels of 'APX', 80.8 ± 1.00 and 87.93 ± 0.38 μmol $min^{-1}$ mg $^{-1}$ protein of fresh weight as seen by the activity profile of 'APX' (Figure 12.6) at the end of 14-day experiment in 60 mol $m^{-3}$ and 120 mol $m^{-3}$ NaCl treatment, whereas IET 4786 decreased with increasing salinity stress 70.07 ± 0.03 and 68.63 ± 0.32 μmol $min^{-1}$ $mg^{-1}$ protein of fresh weight, respectively. Even while 'APX' activity increased in 'IET 4786' under salt stress up to 120 mol $m^{-3}$, it was still much lower than in the unstressed leaves of '*Nona Bokra*' (Tables 12.1 and 12.2). Seed vigour index (SVI) and germination percentages were calculated and values represented in Table 12.3. The SVI for 'IET 4786' was found to decrease with the increasing concentration of NaCl. However, the SVI was consistent for var. '*Nona Bokra*' under the NaCl concentration of 0 and 60 mol $m^{-3}$.

'Proline' accumulation was already evident in both cultivars, although it was consistently higher in the salt-sensitive cultivar 'IET 4786' than the salt-tolerant '*Nona Bokra*' after 11 and 14 days of stress treatment ($p < 0.05$) (Figure 12.7). In comparison to roots, leaves have a higher content of 'proline' ($p < 0.05$) (Figure 12.7).

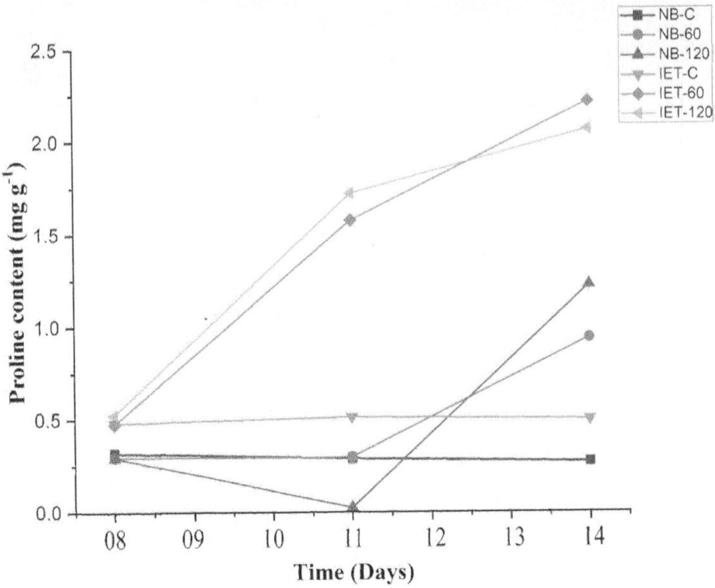


**FIGURE 12.6** Proline content in roots of var. 'IET 4786' and var. 'GD 98028'. Here, NB-C = GD 98028 at control condition, NB-60 = GD 98028 at 60 mol m$^{-3}$ NaCl, NB-120 = GD 98028 at 120 mol m$^{-3}$ NaCl, IET C = IET 4786 at control, IET 60 = IET 4786 at 60 mol m$^{-3}$ NaCl, IET 120 = IET 4786 at 120 mol m$^{-3}$ NaCl.

**TABLE 12.1**
**Two-Way ANOVA Analysis of Variance of a Completely Randomized Design (Two-Way Analysis of Variance), Where the Varieties and Treatments (NaCl Concentrations) and Their Interaction at Day 0**

| | | Statistics | | | |
|---|---|---|---|---|---|
| **Variable** | **Source of Variation** | **Mean Square** | **F-Value** | **P-Value** | **$R^2$ Adj.** |
| SOD | Variety | 298.250 | 167.118 | <0.05 | 0.968 |
| | Treatment | 306.004 | 171.463 | <0.05 | |
| | Variety Treatment | 8.750 | 4.903 | <0.05 | |
| CAT | Variety | 0.000 | 0.214 | >0.05 | 0.466 |
| | Treatment | 0.008 | 8.310 | <0.05 | |
| | Variety Treatment | 0.001 | 1.500 | >0.05 | |
| AP | Variety | 9.976 | 14.791 | <0.05 | 0.565 |
| | Treatment | 2.672 | 3.961 | <0.05 | |
| | Variety Treatment | 1.474 | 2.185 | >0.05 | |

**TABLE 12.2**
**Two-Way ANOVA Analysis of Variance of a Completely Randomized Design (Two-Way Analysis of Variance) Where the Varieties and Treatments (NaCl Concentrations) and Their Interaction Is Represented at Day 6**

| Variable | Source of Variation | Statistics | | | |
|---|---|---|---|---|---|
| | | Mean Square | F-Value | P-Value | $R^2$ Adj. |
| SOD | Variety | 298.250 | 167.118 | <0.05 | 0.977 |
| | Treatment | 306.004 | 171.463 | <0.05 | |
| | Variety Treatment | 8.750 | 4.903 | <0.05 | |
| CAT | Variety | 0.000 | 0.214 | >0.05 | 0.623 |
| | Treatment | 0.008 | 8.310 | <0.05 | |
| | Variety Treatment | 0.001 | 1.500 | >0.05 | |
| AP | Variety | 401.389 | 544.051 | <0.05 | 0.990 |
| | Treatment | 59.067 | 80.061 | <0.05 | |
| | Variety Treatment | 167.244 | 226.686 | <0.05 | |

**TABLE 12.3**
**Seed Vigour Index and Germination Percentages of Var. IET 4786 and Var. GD 98028 Rice Varieties in Different Salt Stress Conditions**

| Salt Stress | Rice Varieties | Seed Vigour Index (SVI) | Germination (%) |
|---|---|---|---|
| **0 mol m$^{-3}$** | GD 98028/*Nona Bokra* | 2.60 ± 0.07 | 87 ± 1.22 |
| | IET 4786 | 2.77 ± 0.03 | 97 ± 2.87 |
| **60 mol m$^{-3}$** | GD 98028/*Nona Bokra* | 2.60 ± 0.10 | 88 ± 1.44 |
| | IET 4786 | 2.27 ± 0.22 | 78 ± 3.86 |
| **120 mol m$^{-3}$** | GD 98028/*Nona Bokra* | 2.21 ± 0.08 | 82 ± 1.65 |
| | IET 4786 | 0.26 ± 0.05 | 10 ± 1.54 |

'Proline' levels were highest in the leaves of the salt-sensitive cultivar 'IET 4786' when treated to the intermediate NaCl dose i.e., 60 mol m $^{-3}$ after 14 days of stress, and a little lower at the highest stress treatment at 120 mol m$^{-3}$.

The distribution of salinity of soil was recorded throughout the Sundarbans before and after the cyclonic event 'Aila' has occurred (Figure 12.8).

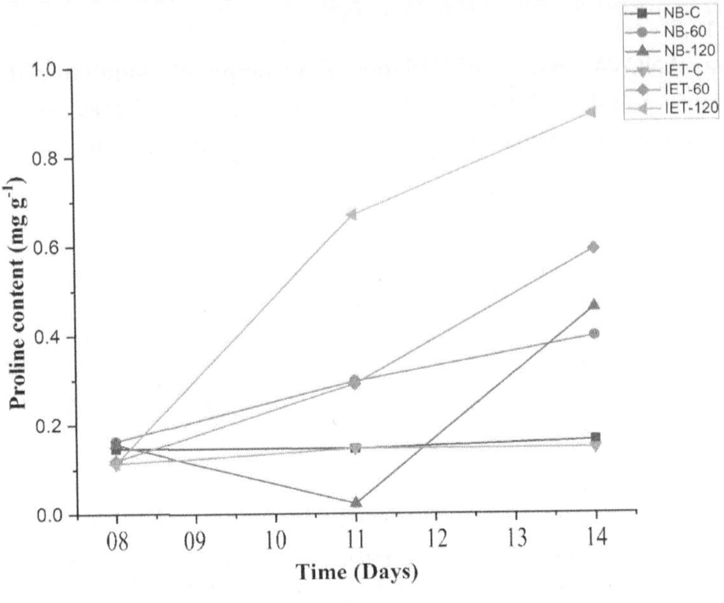


**FIGURE 12.7** Proline content in leaves of var. 'IET 4786' and var. 'GD 98028'. Here, NB-C = GD 98028 at control condition, NB-60 = GD 98028 at 60 mol $m^{-3}$ NaCl, NB-120 = GD 98028 at 120 mol $m^{-3}$ NaCl, IET C = IET 4786 at control, IET 60 = IET 4786 at 60 mol $m^{-3}$ NaCl, IET 120 = IET 4786 at 120 mol $m^{-3}$ NaCl.

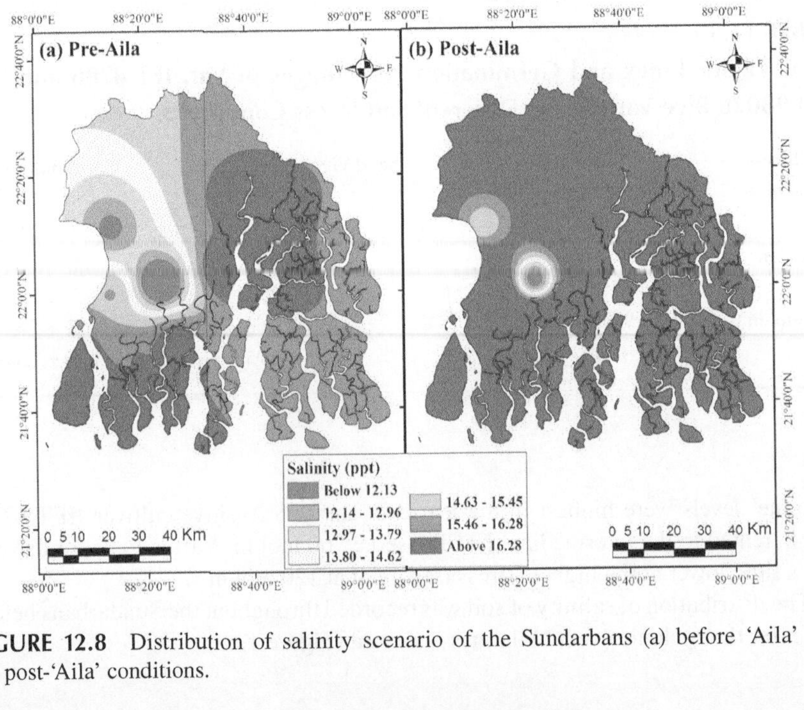


**FIGURE 12.8** Distribution of salinity scenario of the Sundarbans (a) before 'Aila' and (b) post-'Aila' conditions.

## 12.4 DISCUSSION

Salt stress affects almost all physiological functions of the plant. Plants subjected to such environmental stress conditions often suffer oxidative damage due to the disbalance between the production of ROS such as superoxide radical, hydrogen peroxide and hydroxyl radical and the reciprocating action of the antioxidants.

Mohammed and Sen (1990) hypothesized that osmotic stress or particular ion toxicity cause germination to be decreased in saline media. The reverse of this reaction occurs when seeds are translocated from a 'saline water' to 'distilled water' (Sohn et al., 2005). In this study, '*Nona Bokra*' exhibited superior germination during the salinity treatment period than 'IET 4786', suggesting that the latter is more resilient to the stress caused by NaCl. It is commonly believed that a plant's ability to control or block the entry of Na ions from the roots to the leaves determines how resistant it is to salinity (Sohn et al., 2005). The nutritional balance and osmotic control are frequently upset by greater sodium content, which results in specific ion toxicity.

Superoxide radical and hydrogen peroxide production levels in salinized conditions are the best indicators of the severity of oxidative stress. Enzymes of the antioxidant system was investigated in the leaves of salt-sensitive and salt-tolerant cultivars to ascertain the function of ROS scavenging systems in preventing oxidative stress. These present findings regarding 'SOD' activity concur with those previously reported in '*Solanum tuberosum*' (Benavídes et al., 2000), '*Brassica juncea*' (Kumar, 2002), and '*Najas graminea*' (Rout and Shaw, 2001), where sensitive plants displayed lower basal levels of the enzyme activity than did tolerance cultivars. 'SOD' activity increased in salt-tolerant cultivars of wheat (Mandhania et al., 2006), cotton (Meloni et al., 2003), and '*Catharanthus roseus*' whereas it decreased in salt-sensitive cultivars, which is similar to the current findings (Jaleel, 2009). However, it has also been observed that salinity boosts 'SOD' activity in both salt-tolerant and salt-sensitive cultivars of *B. juncea*, with a larger basal level in the salt-tolerant cultivar (Kumar et al., 2006). Increased 'SOD' activity was observed for all the resistant cultivars withstanding 'oxidative stress' (Desingh and Kanagaraj, 2007). In 'salt-sensitive' cultivars, a decline in 'SOD' activity might supress the ability to combat 'oxidative stress'. Inferring a potential role for an enzyme when exposed to salinity, 'SOD' activity increases and by the high basal activity in '*Nona Bokra*' (Rout and Shaw, 2001).

The 'CAT' is a detoxifying enzyme that breaks down hydrogen peroxide into water and oxygen. The results of this study support the findings of Pal et al. (2004) and Mutlu et al. (2009), where, enhanced 'CAT' activity was observed in the 'salt-tolerant' and 'salt-sensitive' cultivars of rice and wheat, respectively. According to Azooz et al. (2009), salt-tolerant types of maize revealed a progressive increase in 'CAT' surge with increased saline stress, but salt-sensitive cultivars showed a considerable decrease. When aquatic macrophytes were subjected to 200 mol $m^{-3}$ NaCl stress, Rout and Shaw (2001) and Dolatabadian et al. (2008) similarly noted an increase in 'CAT' activity. A directly proportional relationship exist between 'CAT' activity and $H_2O_2$ formation in saline circumstances (Gupta and Gupta, 2005; Gueta-Dahan et al., 1997). According to study results, salt-tolerant genotypes of chickpea showed considerably larger 'CAT' surge compared to genotypes that are sensitive both before and after flowering (Singh et al., 2005), helping these genotypes to detoxify $H_2O_2$ generated under saline circumstances. High 'CAT' activity was found in the cultivars of rice under saline soils. When compared to IR 29, a salinity susceptible variety, '*Pokkali*' (a salinity tolerant variety) produced results that were identical to the '*Nona Bokra*' from the present findings.

Benavídes et al. (2000) showed a higher basal level of 'APX' in salt-tolerant potato clones than in salt-sensitive clones, which is consistent with these findings. In a similar manner, Meneguzzo et al. (1999) found that under NaCl stress, salt-sensitive wheat cultivars increased their 'APX' activity significantly more than salt-tolerant wheat cultivars. Rout and Shaw (2001) observed high level of 'APX' activity in 'salt-sensitive' 'aquatic macrophytes' in response to NaCl treatment, and decreased activity in resistant cultivars. Contrarily, Chawla et al. (2013) showed a greater increase in activity in the leaves of the maize genotype that is salt-tolerant than salt-sensitive. 'Wheat' (Heidari and Mesri, 2008; Khan, 2003), 'cotton' (Desingh and Kanagaraj, 2007), and a halophyte '*Cakile maritima*' (Amor et al., 2007) also showed an increase in 'APX' activity with an increase in salt-stress, indicating that a high basal level of 'APX' and salt-induced increase in 'APX' surge may be able to confer tolerance by detoxifying $H_2O_2$ produced when plants are exposed to saline conditions.

One of the most notable adaptation changes observed in numerous plants in response to a wide range of biotic/abiotic stressors is the accumulation of 'proline' (Khare et al., 2020). The concentrations of 'proline' also accumulated less in the salt-resistant cultivar 'IET 4786', indicating that osmotic adjustment may be a factor in stress injury rather than a component of 'proline' build-up as a result of salt resistance (Lutts et al., 1999). Due to its critical functions as a metal chelator, antioxidant, suitable osmolyte, and signalling molecule, higher cellular levels of 'proline' are frequently associated with greater plant performance under stress circumstances (Nahar et al., 2016). Reduced 'proline' breakdown, de novo synthesis, or both may be responsible for stress-responsive 'proline' build-up (dos Reis et al., 2012). Moreover, the present results are in line with the earlier findings of Lutts et al. (1996).

Hence, it could be said that the var. IET 4786 could be promoted to the moderate saline zones of Sundarbans (post-Aila situation of soil, Figure 12.8), which was not practised earlier in this region. The study found that the var. IET 4786 was cultivated in the regions where the salinity was below 13 ppt. However post-Aila changes, forced the agricultural soils towards more saline zones, where var. IET 4786 qualifies to be a better option beside the cultivar *Nona Bokra*. In the present study, the var. IET 4786 was highlighted as it has good market demand. However, studies on biochemical properties of other salinity-tolerant varieties good give different results and definitely an uncertainty related to the present research. The experiments done in the research were continued till 15 days in the environmental chamber to observe the germination and growth pattern but the real field conditions may vary from seasons to year resulting in the variation of results.

## 12.5 IMPLICATIONS FOR ACADEMICIANS AND IMPLICATIONS FOR PRACTITIONERS

This research reveals the physiological and biochemical principles of salt-tolerant rice cultivar. The salt-tolerant 'Nona Bokra' and salt-sensitive 'IET 4786' differ in proline build-up, antioxidant enzyme activity, and germination rates, which may reveal plant stress physiology. With this data, scientists may study the genetic and molecular roots of salt-tolerant variety and create more resistant rice varieties. The study also shows that stress enzymes (SOD, CAT, and APX) and osmo-protectants (proline) are important in analysing how plants react to abiotic stress, paving the way for plant biochemistry and stress adaption research.

Farmers, especially in coastal locations like the Sundarbans, might utilize the study's findings to choose and grow salt-tolerant rice varieties. Since 'Nona Bokra' tolerates high saline conditions, it may be a viable crop in high salinity areas, which is already cultivated. In recent past Sundarbans witnessed frequent mild and severe cyclonic storms. Global climate change leads to rise in sea water level and Sundarbans being a coastal island face the challenge of salt water intrusion. These environmental catastrophes lead to increase salinity in the upward land of Sundarbans. In those area traditional salt-sensitive rice varieties are predominantly cultivated and to cope up with this changing scenario, adaption of the salt-tolerant variety ('Nona Bokra') is difficult than focus on existing salt-sensitive variety ('IET 4786') which can sustain in mild to moderate saline areas. Knowledge of proline and antioxidant enzymes' functions in plant stress responses may impact crop stress tolerance management tactics. Strategies that maintain or boost enzyme activity may improve crop resilience.

## 12.6 CONCLUSIONS

The findings of this experiment showed significant variations between the germination and growth responses of two rice cultivars. The results for the antioxidant enzymes 'SOD', 'APX' and 'CAT's constitutive and salt-induced activity' also suggest that they did not significantly contribute to *Nona Bokra*'s acquisition of salt tolerance. The much larger amount of stress-induced 'proline' accumulation in 'IET 4786' also made it abundantly evident that plants exposed to NaCl have higher levels of that amino acid. '*Nona Bokra*' seeds also showed a higher 'proline' content by virtue of their higher germination rate. Additionally, 'proline' concentrations increased less in the cultivar '*Nona Bokra*', suggesting that osmotic adjustment may be a contributing factor to stress injury rather than a by-product of 'proline' development as a result of salt resistance. Therefore, the rice cultivar 'IET 4786' could be recommended for cultivation under mild to moderate salt-stressed conditions in Sundarbans ecosystem, which is not reported till date.

### 12.6.1 Sustainable Development Implications

This study reveals that 'IET 4786' is sustainable in mild to moderate salt stress. To find out any other such varieties extensive trial and error-based research is required which is an uphill struggle. Identification of such varieties using machine learning-based predictive modelling with the features of identified salt-sensitive varieties that can sustain in increase salinity may be highly useful as well as cost- and time-effective solution. Such artificial intelligence (AI)-based models can predict agricultural productivity following environmental changes. Government, scientist and other related authorities can use these models to develop sustainable agricultural practices and management to withstand climate change and other stressors.

Large datasets from experiments like this may help AI-driven systems to predict rice cultivar performance under varied stress conditions. This may help to identify the optimum types for certain locations, enhancing agricultural production and reducing resource waste.

To ensure global food security in the changing climatic situation, more salt-tolerant rice varieties are required. Besides finding among the existing varieties, some new genetically modified varieties are also required which will be highly salt tolerant as well as productive. ML-based predictive model can suggest optimum combinations of genetic-based characteristics features that can accelerates invention of new varieties.

## ACKNOWLEDGEMENTS

The work is funded by DST-SERB, Government of India, New Delhi, Project EMR/2016/002618.

## REFERENCES

Allakhverdiev, S.I., Nishiyama, Y., Suzuki, I., Tasaka, Y. and Murata, N. (1999). Genetic engineering of the unsaturation of fatty acids in membrane lipids alters the tolerance of *Synechocystis* to salt stress. *Proceedings of the National Academy of Sciences*, *96*(10), pp. 5862–5867.

Amor, N.B., Jiménez, A., Megdiche, W., Lundqvist, M., Sevilla, F. and Abdelly, C. (2007). Kinetics of the anti-oxidant response to salinity in the halophyte *Cakile maritima*. *Journal of Integrative Plant Biology*, *49*(7), pp. 982–992.

Awaad, H.A., 2023. Introduction to "Salinity Resilience and Sustainable Crop Production under Climate Change". In *Salinity Resilience and Sustainable Crop Production under Climate Change* (pp. 3–25). Cham: Springer Nature Switzerland.

Azooz, M.M., Ismail, A.M. and Elhamd, M.A. (2009). Growth, lipid peroxidation and antioxidant enzyme activities as a selection criterion for the salt tolerance of maize cultivars grown under salinity stress. *International Journal of Agriculture and Biology*, *11*(1), pp. 21–26.

Basu, S., Roychoudhury, A., Saha, P.P. and Sengupta, D.N. (2010). Differential antioxidative responses of indica rice cultivars to drought stress. *Plant Growth Regulation*, *60*(1), pp. 51–59.

Benavídes, M.P., Marconi, P.L., Gallego, S.M., Comba, M.E. and Tomaro, M.L. (2000). Relationship between antioxidant defence systems and salt tolerance in *Solanum tuberosum*. *Functional Plant Biology*, *27*(3), pp. 273–278.

Blume, D.E. and McClure, J.W. (1980). Developmental effects of Sandoz 6706 on activities of enzymes of phenolic and general metabolism in barley shoots grown in the dark or under low or high intensity light. *Plant Physiology*, *65*(2), pp. 238–244.

Bor, M., Özdemir, F. and Türkan, I. (2003). The effect of salt stress on lipid peroxidation and antioxidants in leaves of sugar beet *Beta vulgaris* L. and wild beet *Beta maritima* L. *Plant Science*, *164*(1), pp. 77–84.

Cao, H., Ding, R., Kang, S., Du, T., Tong, L., Zhang, Y., Chen, J. and Shukla, M.K. (2023). Drought, salt, and combined stresses in plants: Effects, tolerance mechanisms, and strategies. *Advances in Agronomy*, *178*, pp. 107–163.

Chawla, S., Jain, S. and Jain, V. (2013). Salinity induced oxidative stress and antioxidant system in salt-tolerant and salt-sensitive cultivars of rice (*Oryza sativa* L.). *Journal of Plant Biochemistry and Biotechnology*, *22*(1), pp. 27–34.

Choudhary, J.R., Ganie, S.A., Bansal, M., Goswami, B., Burdak, A., Tripathi, A. and Wani, S.H. (2022). Emerging Roles of Osmoprotectants in the Abiotic Stress Tolerance of Plants. In *Plant Abiotic Stress Physiology* (pp. 263–287). Apple Academic Press, US and Canada.

Chowdhury, A., Naz, A., Sharma, S.B. and Dasgupta, R. (2023). Changes in salinity, mangrove community ecology, and organic blue carbon stock in response to cyclones at Indian Sundarbans. *Life*, *13*(7), p. 1539.

Demiral, T. and Türkan, I. (2005). Comparative lipid peroxidation, antioxidant defense systems and proline content in roots of two rice cultivars differing in salt tolerance. *Environmental and Experimental Botany*, *53*(3), pp. 247–257.

Desingh, R. and Kanagaraj, G. (2007). Influence of salinity stress on photosynthesis and antioxidative systems in two cotton varieties. *General and Applied Plant Physiology*, *33*(3–4), pp. 221–234.

Dolatabadian, A., Sanavy, S.M. and Chashmi, N.A. (2008). The effects of foliar application of ascorbic acid (vitamin C) on antioxidant enzymes activities, lipid peroxidation and "proline" accumulation of canola (*Brassica napus* L.) under conditions of salt stress. *Journal of Agronomy and Crop Science*, *194*(3), pp. 206–213.

dos Reis, S.P., Lima, A.M. and De Souza, C.R.B. (2012). Recent molecular advances on downstream plant responses to abiotic stress. *International Journal of Molecular Sciences*, *13*(7), pp. 8628–8647.

Dubey, R.S. and Rani, M. (1989). Salinity induced accumulation of free amino acids in germinating rice seeds differing in salt tolerance. *Journal of Agronomy and Crop Science*, *163*(4), pp. 236–247.

Egea, I., Estrada, Y., Faura, C., Egea-Fernández, J.M., Bolarin, M.C. and Flores, F.B. (2023). Salt-tolerant alternative crops as sources of quality food to mitigate the negative impact of salinity on agricultural production. *Frontiers in Plant Science*, *14*, p.1092885.

FAO, 2017. The Food and Agriculture Organization of the United Nations. World Hunger on the Rise. http://www.fao.org/state-of-food-security-nutrition.

Gueta-Dahan, Y., Yaniv, Z., Zilinskas, B.A. and Ben-Hayyim, G. (1997). Salt and oxidative stress: Similar and specific responses and their relation to salt tolerance in citrus. *Planta*, *203*(4), pp. 460–469.

Gul, Z., Tang, Z.H., Arif, M. and Ye, Z. (2022). An insight into abiotic stress and influx tolerance mechanisms in plants to cope in saline environments. *Biology*, *11*(4), p. 597.

Guo, J., Lu, X., Tao, Y., Guo, H. and Min, W. (2022). Comparative ionomics and metabolic responses and adaptive strategies of cotton to salt and alkali stress. *Frontiers in Plant Science*, *13*, p.871387.

Gupta, S. and Gupta, N.K. (2005). High temperature induced antioxidative defense mechanism in seedlings of contrasting wheat genotypes. *Indian Journal of Plant Physiology*, *10*(1), p.73.

Heidari, M. and Mesri, F. (2008). Salinity effects on compatible solutes, antioxidants enzymes and ion content in three wheat cultivars. *Pakistan Journal of Biological Sciences: PJBS*, *11*(10), pp. 1385–1389.

Jaleel, C.A. (2009). Soil salinity regimes alters antioxidant enzyme activities in two varieties of 'CAT' haranthus roseus. *Botany Research International*, *2*(2), pp. 64–68.

Jayabalan, S., Rajakani, R., Kumari, K., Pulipati, S., Hariharan, R.V.G., Venkatesan, S.D., Jaganathan, D., Kancharla, P.K., Raju, K. and Venkataraman, G. (2022). Morpho-physiological, biochemical and molecular characterization of coastal rice landraces to identify novel genetic sources of salinity tolerance. *Plant Physiology and Biochemistry*, *187*, pp. 50–66.

Khan, N.A. (2003). NaCl-inhibited chlorophyll synthesis and associated changes in ethylene evolution and antioxidative enzyme activities in wheat. *Biologia Plantarum*, *47*(3), pp. 437–440.

Khare, T., Srivastava, A.K., Suprasanna, P. and Kumar, V. (2020). Individual and additive stress impacts of $Na^+$ and $Cl^-$ on "proline" metabolism and nitrosative responses in rice. *Plant Physiology and Biochemistry*, *152*, pp. 44–52.

Kumar, M., 2002. *Effect of Sodium Chloride on Activated Oxygen-Scavenging Enzymes in Salt Tolerant and Salt-Sensitive Cultivars of Brassica Juncea L.* (Doctoral dissertation, Biochemistry, CCSHAU, Hisar).

Kumar, M., Jain, S. and Jain, V. (2006). Effect of NaCl stress on osmotic adjustment, ionic homeostasis and yield attributes in salt-sensitive and salt-resistant cultivars of *Brassica juncea* L. *Physiology and Molecular Biology of Plants*, *12*(1), p. 75.

Laamari, I., Schnug, E., Béjaoui, Z., Slaimi, R. and Abassi, M. (2024). Differential responses of six *Casuarina* species to NaCl seed priming and salt stress during germination and early seedling growth. *Arid Land Research and Management*, *38*4), pp. 1–13.

Lutts, S., Kinet, J.M. and Bouharmont, J. (1996). Effects of salt stress on growth, mineral nutrition and "proline" accumulation in relation to osmotic adjustment in rice (*Oryza sativa* L.) cultivars differing in salinity resistance. *Plant Growth Regulation*, *19*(3), pp. 207–218.

Lutts, S., Majerus, V. and Kinet, J.M. (1999). NaCl effects on "proline" metabolism in rice (*Oryza sativa)* seedlings. *Physiologia Plantarum*, *105*(3), pp. 450–458.

Mandal, P., Roy, P., Maiti, A., Yadav, A., Bhattacharya, S. and Paul, S. (2024). Assessing the interplay of environmental change, cyclonic hazards, and their ramifications on agriculture in Indian Sundarban. *Ocean & Coastal Management*, *251*, p. 107080.

Mandhania, S., Madan, S. and Sawhney, V. (2006). Antioxidant defense mechanism under salt stress in wheat seedlings. *Biologia Plantarum*, *50*(2), pp. 227–231.

McCord, J.M. and Fridovich, I. (1969). Superoxide dismutase: An enzymic function for erythrocuprein (hemocuprein). *Journal of Biological Chemistry*, *244*(22), pp. 6049–6055.

McNeil, S.D., Nuccio, M.L. and Hanson, A.D. (1999). Betaines and related osmoprotectants. Targets for metabolic engineering of stress resistance. *Plant Physiology*, *120*(4), pp. 945–949.

Meloni, D.A., Oliva, M.A., Martinez, C.A. and Cambraia, J. (2003). Photosynthesis and activity of superoxide dismutase, peroxidase and glutathione reductase in cotton under salt stress. *Environmental and Experimental Botany*, *49*(1), pp. 69–76.

Meneguzzo, S., Navam-Izzo, F. and Izzo, R. (1999). Antioxidative responses of shoots and roots of wheat to increasing NaCl concentrations. *Journal of Plant Physiology*, *155*(2), pp. 274–280.

Mishra, A.K., Das, R., George Kerry, R., Biswal, B., Sinha, T., Sharma, S., Arora, P. and Kumar, M. (2023). Promising management strategies to improve crop sustainability and to amend soil salinity. *Frontiers in Environmental Science*, *10*, p. 962581.

Mistry, A.H., 2022. Soil Salinity Effects on Traditional Agricultural Practices in Three Coastal Rural Villages of Indian Sundarban, West Bengal. In *Transforming Coastal Zone for Sustainable Food and Income Security* (pp. 1043–1058). Cham: Springer.

Mohammed, S. and Sen, D.N. (1990). Germination behavior of some halophytes in Indian desert. *Indian Journal of Experimental Biology*, *28*(6), pp. 545–549.

Munns, R. (2002). Comparative physiology of salt and water stress. *Plant, Cell & Environment*, *25*(2), pp. 239–250.

Mutlu, S.A.L.I.H., Atici, Ö. and Nalbantoglu, B. (2009). Effects of salicylic acid and salinity on apoplastic antioxidant enzymes in two wheat cultivars differing in salt tolerance. *Biologia Plantarum*, *53*(2), pp. 334–338.

Nahar, K., Hasanuzzaman, M. and Fujita, M., 2016. Roles of Osmolytes in Plant Adaptation to Drought and Salinity. In *Osmolytes and Plants Acclimation to Changing Environment: Emerging Omics Technologies* (pp. 37–68). New Delhi: Springer India.

Nakano, Y. and Asada, K. (1981). Hydrogen peroxide is scavenged by ascorbate-specific peroxidase in spinach chloroplasts. *Plant and Cell Physiology*, *22*(5), pp. 867–880.

Negacz, K., Malek, Ž, de Vos, A. and Vellinga, P. (2022). Saline soils worldwide: Identifying the most promising areas for saline agriculture. *Journal of Arid Environments, 203*, p. 104775.

Omar, M.M., Shitindi, M.J., Massawe, B.J., Fue, K.G., Meliyo, J.L. and Pedersen, O. (2023). Salt-affected soils in Tanzanian agricultural lands: Type of soils and extent of the problem. *Sustainable Environment, 9*(1), p. 2205731.

Ondrasek, G., Rathod, S., Manohara, K.K., Gireesh, C., Anantha, M.S., Sakhare, A.S., Parmar, B., Yadav, B.K., Bandumula, N., Raihan, F. and Zielińska-Chmielewska, A. (2022). Salt stress in plants and mitigation approaches. *Plants, 11*(6), p. 717.

Pal, M., Singh, D.K., Rao, L.S. and Singh, K.P. (2004). Photosynthetic characteristics and activity of antioxidant enzymes in salinity tolerant and sensitive rice cultivars. *Indian Journal of Plant Physiology, 9*(4), pp. 407–412.

Pamungkas, S.S.T. and Farid, N. (2022). Drought stress: Responses and mechanism in plants. *Reviews in Agricultural Science, 10*, pp. 168–185.

Pathak, P.S. (2000). Agro forestry: A tool for arresting land degradation. *Indian Farming, 49*(11), pp. 15–19.

Rashid, A., Schutte, B.J., Ulery, A., Deyholos, M.K., Sanogo, S., Lehnhoff, E.A. and Beck, L. (2023). Heavy metal contamination in agricultural soil: Environmental pollutants affecting crop health. *Agronomy, 13*(6), p. 1521.

Rhodes, D., Verslues, P.E. and Sharp, R.E. (1998). Role or Amino Acids in Abiotic Stress Resistance. In *Plant Amino Acids* (pp. 333–370). Boca Raton, FL: CRC Press.

Rout, N.P. and Shaw, B.P. (2001). Salt tolerance in aquatic macrophytes: Possible involvement of the antioxidative enzymes. *Plant Science, 160*(3), pp. 415–423.

Roy, S.S. and Ghosh, T. (2024). Local-level impacts of Cyclone Yaas on the Islands of the Indian Sundarbans Delta. *Natural Hazards, 120*(4), pp.3995–4010.

Sachdev, S., Ansari, S.A. and Ansari, M.I. (2023). Reactive Oxygen Species in Plants. Springer, Singapore.

Seleiman, M.F., Aslam, M.T., Alhammad, B.A., Hassan, M.U., Maqbool, R., Chattha, M.U., Khan, I., Gitari, H.I., Uslu, O.S., Roy, R. and Battaglia, M.L. (2022). Salinity stress in wheat: Effects, mechanisms and management strategies. *Phyton – International Journal of Experimental Botany, 91*(4), pp. 0031–9457.

Singh, A., Mehta, S., Yadav, S., Nagar, G., Ghosh, R., Roy, A., Chakraborty, A. and Singh, I.K. (2022). How to cope with the challenges of environmental stresses in the era of global climate change: An update on ROS stave off in plants. *International Journal of Molecular Sciences, 23*(4), p. 1995.

Singh, R.A., Singh, A.P., Roy, N.K. and Singh, A.K. (2005). Pigment concentration and activity of antioxidant enzymes in zinc tolerant and susceptible chickpea genotypes subjected to zinc stress. *Indian Journal of Plant Physiology, 10*(1), pp. 48–53.

Sohn, Y.G., Lee, B.H., Kang, K.Y. and Lee, J.J. (2005). Effects of NaCl stress on germination, antioxidant responses, and "proline" content in two rice cultivars. *Journal of Plant Biology, 48*(2), pp. 201–208.

Tenhunen, S., Uddin, M.J. and Roy, D. (2023). After cyclone Aila: Politics of climate change in Sundarbans. *Contemporary South Asia, 31*(2), pp.222–235.

Upadhyay, S.K., Singh, J.S. and Singh, D.P. (2011). Exopolysaccharide-producing plant growth-promoting rhizobacteria under salinity condition. *Pedosphere, 21*(2), pp. 214–222.

Wu, J., Nadeem, M., Galagedara, L., Thomas, R. and Cheema, M. (2022). Effects of chilling stress on morphological, physiological, and biochemical attributes of silage corn genotypes during seedling establishment. *Plants, 11*(9), p .1217.

Zeng, L. and Shannon, M.C. (2000). Salinity effects on seedling growth and yield components of rice. *Crop Science, 40*(4), pp. 996–1003.

# Index

## O



## P



## R



## S



## T



## U



## V



## W